Thank you to my committee Christine Young, Evelyn Maizels, Kevin Brennan, Leah Lebowicz for their aid, guidance, and support in this project.

Thank you to John Daugherty for his leadership and the UIC Biomedical Visualization program for the opportunity to explore and learn about these fascinating proteins.

Thank you to Mom and Dad for all the help and support you have given me in this endeavor and all other ones.

CONTENTS

INTRODUCTION: 4

GPCR ESSENTIALS 4
   I The Science of GPCRs: 4
   II Biochemistry Behind GPCRs 6
   III Ligand Binding and Functional Selectivity 7
   IV Dynamic Helices: 8
   V Allosteric and Orthosteric 9
   VI GPCRs Forming Dimers and Oligomers 10

THE GUIDE FOR BEST PRACTICES:
THE ACCURATE DEPICTION OF GPCRS 12
   I When to use 2D or 3D Styles 12
   II Purpose and Audience 12
   III Designing the Stage 12
   IV Stylistic Choices 13
   V Challenges and Pitfalls 14
   VI 2D Best Practices 15
   VII Databases 16
   VIII 3D Representations 19

CONCLUDING REMARKS 24

ACKNOWLEDGEMENTS 24

WORKS CITED REFERENCES 24

APPENDIX A. GLOSSARY OF TERMS RELEVANT
TO G PROTEIN-COUPLED RECEPTORS 27

APPENDIX B. G PROTEIN SIGNALING 31

APPENDIX C. CHART OF AMINO ACIDS 34

INTRODUCTION:

G Protein-Coupled Receptors (GPCRs) are transmembrane proteins. They play an important role in cell signaling and consequently they play an important role in pharmacology. Scientists are working to use GPCRs for many different reasons, and as the practical knowledge of GPCRs increases the need for accurate illustrations has become urgent. Since the pharmacology of GPCRs is very prominent in current research it is extremely important that professional and student medical illustrators understand the science of GPCRs. Medical illustrators' understanding of the science of GPCRs will create better illustrations and animations which will be a driving force furthering scientific discovery. The medical illustration community needs a guide that explains GPCRs in a way that caters specifically to the needs of medical illustrators. This resource discusses the unique structure, function, and conformational flexibility of GPCRs, what makes them so important in pharmacology, and what makes them so difficult to depict correctly. This guidebook will explain the potential problems that one would encounter when illustrating GPCRs, as well as the solutions to those problems in order to create the most correct and scientifically accurate depictions of GPCRs.

This information can be used to promote further discovery in the field of GPCRs as well as creating a better understanding of information between researchers, clinicians, and patients. This guibebook is divided into two different sections. The first section will explain the important scientific details of GPCRs including conserved structural motifs, ligand binding, allosteric and orthosteric sites, and the oligomerization of GPCRs. The second section of the guidebook describes the conventions for drawing GPCRs, when to use realistic or iconic depictions, and when to use 2D or 3D representations. This section will also take time to discuss how to use the Protein Data Bank (PDB) in order to create accurate illustrations of GPCRs. This guidebook will allow medical illustrators to generate more effective images of GPCRs and serve as a comprehensive resource for illustrating GPCRs.

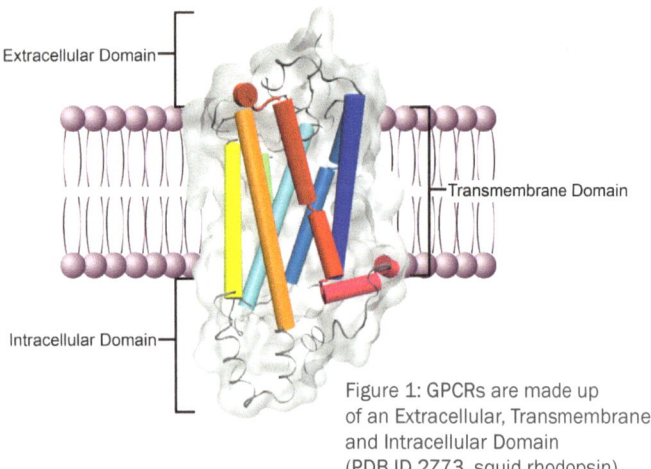

Figure 1: GPCRs are made up of an Extracellular, Transmembrane, and Intracellular Domain (PDB ID 2Z73, squid rhodopsin).

GPCR ESSENTIALS

I THE SCIENCE OF GPCRS:

G-Protein Coupled Receptors (GPCRs) are transmembrane proteins. They represent a versatile family of plasma membrane receptors that affect downstream signaling. They act as cell surface receptors, recognizing and responding to external stimuli by initiating intracellular signaling events. The external stimuli are typically ligands, which bind at or near the extracellular face. Ligands are binding partners of receptors, such as GPCRs, and can be any substance that binds to or is recognized by the receptor. The fundamental role GPCRs play in pharmacology has been steadily developing recently especially due to the refined insights into the mechanisms with which they bind with ligands. GPCRs are categorized into four major classes and several subfamilies (Table 1). These classes are determined based on the primary structure of the GPCR (Katritch, Cherezov, & Stevens, 2013). There are a large number of orphan receptors, or GPCRs that have not yet had their ligand identified (Katritch, Cherezov, & Stevens, 2013). The different families and subfamilies of GPCRs play central roles in many physiological processes from sensory, including vision, smell, and taste, to neurological, cardiovascular, endocrine, and reproductive functions.

CLASS	FAMILY NAME	NUMBER OF MEMBERS
A	Rhodopsin-like Receptors	701 members
B*	Secretin Receptors Adhesion Receptors*	15 members 24 members
C	Glutamate Receptors	15 members
F	Frizzled Receptors	24 members

Table 1: Class and Family names for Human GPCRs including the number of members identified in those families (Katritch et al., 2013). *Adhesion Receptor subfamily is sometimes classed as a separate class (Millar & Newton, 2010), and sometimes considered a subfamily under the B class with the Secretin Receptors (Katritch et al., 2013).

GPCRs all share very similar structure including an extracellular domain, transmembrane domain, and an intracellular domain. The transmembrane domain consists of the iconic seven helical spans traversing the membrane in an anti-parallel fashion. The N-terminal extension sequences, and three loops joining the helices at the extracellular face comprise the extracellular domain (Millar & Newton, 2010). The intracellular domain consists of the C-terminal extension sequences and the three loops joining the helices at the intracellular face (Figure 1). They respond to several extracellular stimuli including photons, small molecules, and proteins. GPCRs also have an additional helix eight that holds itself parallel to the phospholipid bi-layer of the cell membrane by a fatty lipid extension (Figure 2). The transmembrane domain is hydrophobic while the intracellular and extracellular domains are hydrophilic. This affects how lipids bind to GPCRs and what implications that has for pharmacology (Chini & Parenti, 2009).

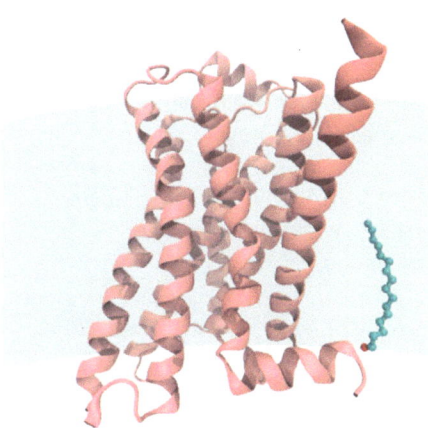

Figure 2: Lipid extension of helix 8 in a Beta2-Adrenergic receptor (PDB ID 2HR1)

A GPCR monomer is comprised of a single polypeptide chain of amino acids. Since different amino acids play different roles in GPCRs the need for consistent nomenclature was addressed. The Ballesteros and Weinstein Nomenclature is the accepted nomenclature for referring to amino acids in the transmembrane domain of a GPCR. This Ballesteros Weinstein convention allows for comparison of closely corresponding residues across a spectrum of GPCR structures of widely different sizes. While the Ballesteros and Weinstein nomenclature is derived from the primary amino acid sequence structure, there is no direct correlation between a particular amino acid's Ballesteros Weinstein number and the residue number assigned to that same amino acid in a conventional primary sequence index, such as the UniProt sequence number or Protein Data Bank (PDB) residue (resid) number (Gonzalez, Cordomi, Matsoukas, Zachmann, & Pardo, 2014; Moreira, 2014).

Within the Ballesteros Weinstein system, each transmembrane residue is assigned two numbers N1 and N2. N1 is the transmembrane helix number (one through seven) and N2 is the number relative to the most conserved, or commonly found, residue in the transmembrane helix. The most conserved amino acid residue is assigned 50, and neighboring amino acid residues are numbered accordingly. These numbers decrease toward the N-terminus and increase towards the C-terminus. For example in helix three, Arginine, is the most conserved amino acid residue so it is referred to as $Arg^{3.50}$ (Gonzalez et al., 2014).

The different helices and loops play import roles in receptor function. GPCRs are arranged in a roughly cylindrical bundle within the membrane. They do not make a perfect cylinder, however, mostly due to the placement of transmembrane helix 3 (TM3). TM3 is the longest of the helices and is notably on a diagonal angle which is due to a conserved cysteine amino acid in TM3 that forms a disulfide bridge to extracellular loop two (ECL2). TM3 typically occupies the space within the bundle of rods. One consequence of the diagonal orientations of helix three is the ionic lock (D[E]RY). On the interior side of the helix contains the conserved structural motif of the ionic lock. On helix three this includes the aspartic ($D^{3.49}$) or glutamic ($E^{3.49}$) as well as the arginine ($R^{3.50}$) and tyrosine ($Y^{3.51}$) together comprise the D[E]RY motif. The arginine on TM3 attaches to the glutamate on helix 6 (TM6) especially in rhodopsin class receptors (Moreira, 2014) (Figure 3).

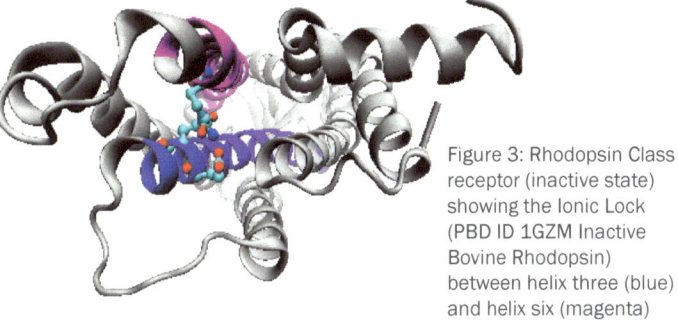

Figure 3: Rhodopsin Class receptor (inactive state) showing the Ionic Lock (PBD ID 1GZM Inactive Bovine Rhodopsin) between helix three (blue) and helix six (magenta)

TM6 not only plays a role in the conserved motif of the ionic lock, but it also features the amino acid conserved motif of the rotamer toggle (CWxxP). The rotamer toggle only occurs in some GPCRs, and in these cases it serves a role in the conformational motility of TM6. TM6 moves around a great deal when a ligand binds to the extracellular surface of the GPCR and activates it. TM6 coupled with helix five (TM5) moves outward in a pendulum like fashion in order to create a larger intracellular binding pocket. The rotamer toggle works due to several aromatic residues that are clustered near a conserved tryptophan ($W^{6.48}$). In the inactive state the cluster of aromatic residues in TM6 points towards helix 7 (TM7). Activation is accompanied by rotation of TM6: the twist motion of TM6 around its longitudinal axis reorients these aromatic residues near the tryptophan so that the cluster points towards TM5 in the active state (Moreira, 2014). The rotation of TM6 works in conjunction with the outward pendulum motion. While this has been proposed and observed in various biophysical methodologies, more research needs to be done in terms of x-ray structure to confirm rotameric movement. As TM6 and TM5 move to open up the intracellular binding pocket, TM7 moves inward towards the core of the GPCR which stabilizes the GPCR in the active conformation. The amino acid residues on TM7 that are responsible for this motion make up the asparagine, proline, xx, tyrosine (NPxxY) conserved motif (Audet & Bouvier, 2012).

The elucidation of x-ray crystallographic structures of GPCRs has been monumental to research in understanding the function and conformational flexibility of GPCRs. The structures have revealed the ligand-binding cavity, which provides access to diffusible small ligands (Jacobson & Costanzi, 2012). Understanding of how ligand binding alters the structure and function of GPCRs to mediate signaling has undergone an expansion in recent years. Concepts such as ligands acting as functionally selective biased agonists to elicit a specific subset of signaling responses are now at the forefront of GPCR research. GPCRs were originally considered to be monomeric, but increasing evidence indicates that they can form dimers and oligomers as well.

II BIOCHEMISTRY BEHIND GPCRS

In biochemistry, proteins are organized into four protein structure levels: primary, secondary, tertiary, and quaternary. The primary sequence is the amino acid sequence. The primary structure for GPCRs would be the amino acid sequence of a polypeptide chain, presented in a linear manner starting with the N-terminus and finishing with the C-terminus. Linear motifs may be presented as features of the primary sequence. This makes it easier to observe

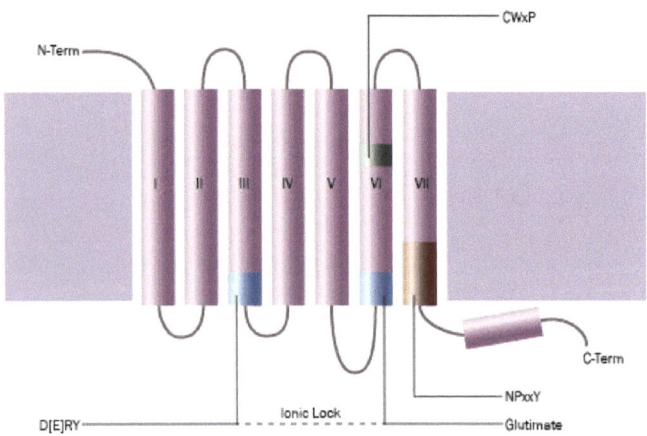

Figure 4: A conventional snake diagram of the most common conserved motifs mapped out on a standard GPCR. Based on diagram 1 in Audet, 2012.

conserved motifs such as the D[E]RY motif (Figure 4). The secondary structure corresponds to localized short-range structural conformations created by propensity of the peptide backbone to be stabilized by hydrogen bonding. Common secondary structural motifs are alpha helices, and beta strands or sheets. Alpha helices are stabilized by backbone hydrogen bonds aligned parallel to the axis of the local structure. The beta strands/sheets are stabilized by backbone hydrogen bonds aligned perpendicular to the local structure. Secondary structural motifs include turns and coils (Millar & Newton, 2010). The predominant secondary structure feature of the GPCR is the alpha helix, which forms each of the seven transmembrane helices. The tertiary structure describes the overall structure of a single polypeptide chain formed by folding short range structural domains into a unique three dimensional shape. The predominant tertiary structure feature of a GPCR corresponds to the arrangement of the GPCR helices to form

the "bundle of rods" in 3D space. The quaternary structure describes the three dimensional arrangement of a protein complex comprised of multiple subunits i.e. multiple polypeptide chains. This defines the unique spatial relationships of the subunits to each other.

Since GPCRs are transmembrane proteins embedded in the cellular phospholipid bi-layer, it is extremely important to understand how they interact with the membrane as well as size relationships. The transmembrane helices typically span the membrane. Each transmembrane helical segment contains roughly twenty-five amino acids. The turns in each helix contains 3.6 amino acids. Each alpha helix contains 5.4 angstroms per complete helical turn. The membrane is roughly forty angstroms in width, corresponding to roughly seven helical turns. Helices five (TM5) and six (TM6), notably, have intracellular extensions. This is very clearly depicted in the solved crystal structure of squid rhodopsin (Figure 5).

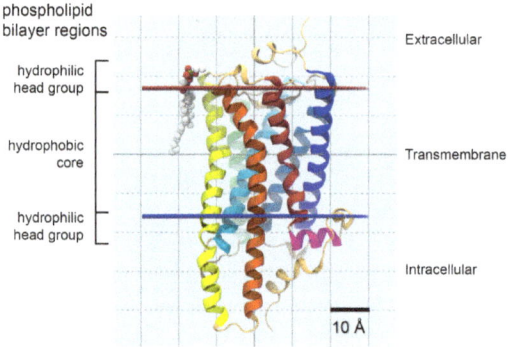

Figure 5: Squid Rhodopsin Receptor (PDB ID 2Z73). A typical membrane is around 40 angstroms long, roughly the length of the transmembrane domain. Red and blue lines denote the hydrophobic core of membrane, see OPM section in text below. Image courtesy of Evelyn Maizels.

Another aspect of GPCRs is their hydrophobicity (Figure 6). Many of the residues that form transmembrane helices are hydrophobic, allowing the transmembrane domain to reside embedded within hydrophobic portion of the lipid bilayer. The intracellular and extracellular loops tend to be more hydrophilic, allowing these structures to reside in contact with the cytoplasmic and extracellular aqueous environments. Within the deep interior of the transmembrane helix bundle however, extended cavities do contain hydrophilic residues, and maintain a structured water network which participates in hydrogen bonding as well as stabilization. Notably the water molecules that make up the structured water network are not part of the bulk aqueous environment. This structured water network supports structural and functional plasticity. The water network plays a role in several GPCRs, e.g. in opioid receptors it functions in stabilizing the transmembrane helices.

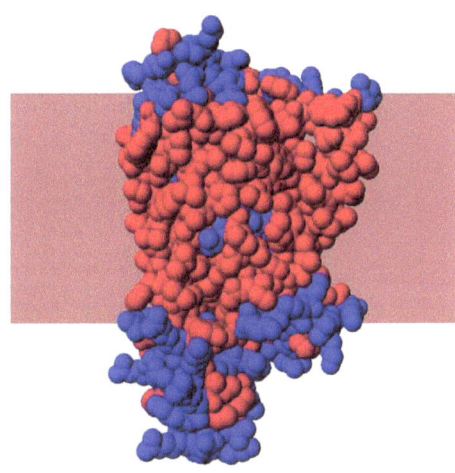

Figure 6: Hydrophobicity of a Turkey Beta1-Adrenergic Receptor (PDB ID 2YCZ). Red indicates hydrophobic residues and blue indicates hydrophilic residues.

III LIGAND BINDING AND FUNCTIONAL SELECTIVITY

GPCRs act through ligand binding. Ligands are molecules that bind to a receptor or enzyme in order to elicit a chemical response. The GPCR superfamily comprises the largest target family for pharmaceutical ligands. It is reported that approximately forty percent of pharmaceuticals currently marketed target GPCRs (Vischer, Watts, Nijmeijer, & Leurs, 2011). GPCRs unique structure and cell surface location make them ideal "druggable" targets for different drug therapies, assuring interest in the pharmaceutical and clinical medicine communities.

Ligand binding for small molecules occurs within ligand binding pockets (Figure 7), which are cavities within the transmembrane domain that generally communicate with the extracellular space. Larger ligands are accommodated within ligand binding sites formed by extracellular loops and N-terminal extension sequences. Ligand binding can induce dynamic rearrangements of the transmembrane helices, which allows them to shift in relation to each other. Subsequently intracellular signaling events, such as downstream signaling cascades initiated by G proteins or by arrestins, and regulatory events such as phosphorylation and desensitization, are initiated within or near the intracellular domain as a consequence of these helical shifts (Whalen, Rajagopal, & Lefkowitz, 2011). GPCR families are defined by what ligands bind to them. Some GPCRs are referred to as orphan receptors because while they have a similar structure to other identified receptors, their endogenous ligand has not yet been identified (Katritch et al., 2013).

Endogenous ligands are natural ligands that originate within the physiological system, and exogenous ligands are ligands that are added to a physiological system, either as a drug in a pharmacological setting, or as an external reagent in an experimental research setting. Exogenous ligands may be isolated from naturally occurring sources. They may also be manufactured as synthetic compounds by research or pharmaceutical laboratories. Exogenous ligands may mimic the action of the endogenous naturally occurring ligand upon a receptor, or may expand the repertoire of responses that the receptor may display (Heng, Aubel, & Fussenegger, 2013).

Ligands induce a variety of responses within the cell, and can be classed according to the type of response. First, we will distinguish agonists and antagonists. Agonists are substances that bind to GPCRs and trigger a signaling response by the receptor. Antagonists, the opposite of agonists, are substances that bind to GPCRs and interfere with the action of the receptor consequently blocking its function. The general classification of antagonist includes both neutral antagonists which do not alter the activity of the basal state, but block the ability of agonists to induce signaling response; and inverse agonists which decrease the basal signaling activity of the receptor. Partial agonists produce a signaling response that is lower than the maximum response achievable for the given signaling system. Partial agonists can antagonize or reduce the effects of full agonists (Andresen, 2011).

Each GPCR conformational state is associated with its own repertoire of signaling behaviors, and one of the unique things about GPCRs is that they can act through functional selectivity. This means that certain GPCR ligands can selectively activate a subset of signaling pathways available to the receptor without activating the rest of the signaling pathways. The types of agonists that work in functional selectivity are known as biased agonists. Biased agonists have the ability to preferentially stabilize specific GPCR conformations at the exclusion of others (Kenakin, 2011). The role of biased agonists in signaling G Proteins and Beta-Arrestins will be discussed below.

All of these ligands bind within or near the extracellular portion of the GPCR. Orthosteric ligand binding sites are where the primary endogenous ligand binds to the GPCR (see also Allosteric Orthosteric section IV, below). Small ligands usually bind to a discrete cavity formed within the interior of

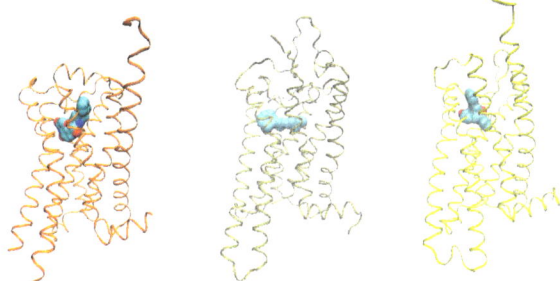

Figure 7: Ligand binding sites in a Beta2-Adrenergic receptor (PDB ID B2AR), Rhodopsin (PDB ID 3PQR), and Opioid Receptor (PDB ID 4N6H).

the helix bundle in the transmembrane region closest to the extracellular domain. Many orthosteric ligands binding sites correspond to this cavity. The shape of the cavity is defined by the placement of transmembrane helix 3 (TM3). TM3 runs diagonally through the core of the GPCR and divides the cavity into two binding pockets. The core binding pocket is where the ligand communicates between TM3 and helix 6 (TM6). The secondary binding pocket communicates between TM3 and helix 7 (TM7) (Figure 8). Larger ligands need more space to bind and as a result the extracellular loops and the N-terminal extensions sequences are recruited to form ligand binding sites for these larger ligands (Venkatakrishnan et al., 2013).

subdivisions of G proteins are the heterotrimeric G proteins and small G proteins. Small G proteins don't communicate directly with GPCRs, but they are activated downstream by diverse signal pathways. Heterotrimeric G proteins are larger biologically active heterotrimers that are comprised of three subunits. They are made up of an alpha, beta, and gamma subunits. GPCRs typically interact with heterotrimeric G proteins directly.

Heterotrimeric G proteins bind to the intracellular surface of the GPCR and the Gα subunit inserts into the hydrophobic pocket of the activated GPCR created by the outward movement of TM6 separating from TM3. The Gα subunit is composed of two domains, the helical domain and the Ras domain. The interaction of the Gα subunit with the GPCR subunit causes the helical domain of the Gα to rotate in relation to the Ras domain of the Gα subunit. The rotation opens up the nucleotide binding pocket to allow the release of the GDP (Preininger, Meiler, & Hamm, 2013). The GTP binds to the Gα subunit resulting in dissociation of the α and βγ subunits from the receptor (Oldham & Hamm, 2008). The subunits then initiate downstream signaling regulating their respective effector proteins which may include enzymes such as adenylyl cyclase (AC) and channels such as the Ca2+ channel (Rasmussen et al., 2011). Some important G alpha proteins include Gαi, Gαq/11, Gαs, and Gα12/13 which will be discussed in Appendix B.

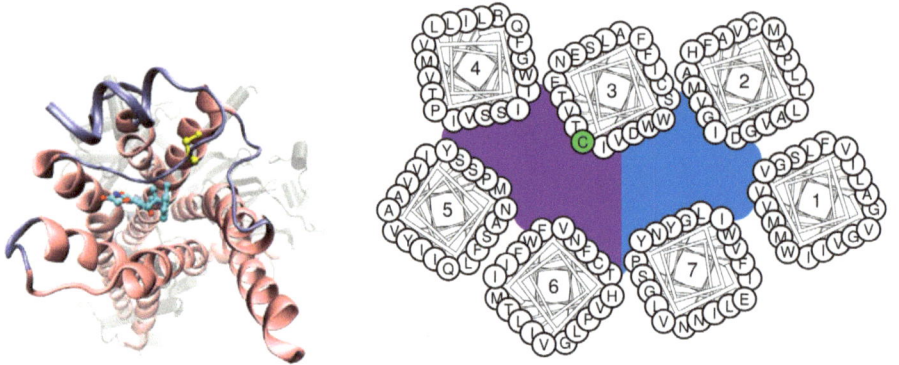

Figure 8: A Beta2-Adrenergic receptor (PDB ID 3SN6) illustrates the division between the major and minor binding pocket both in a 3D diagram and a 2D specialized box plot

Ligand binding causes helical shifts. The shift of the helices relates to the contact the ligand makes within its ligand binding site. This also means that ligands activate different signaling proteins on the intracellular surface. GPCRs were originally thought to only target G Proteins, but now it is known that GPCRs also target arrestins (Violin & Lefkowitz, 2007).

Arrestins are a small family of proteins. They play an important role in regulating signal transduction. This is done by activating or redirecting pathways. Arrestins signal in a positive manner through activation of downstream kinase cascases as described in Appendix B. Arrestins also play a role in receptor desensitization and internalization. Internalization is the uptake of a cell surface component into the interior of the cell. This means that the GPCR is removed from the plasma membrane through endocytosis which is initiated by arrestin binding. Desensitization describes the loss of responsiveness of the signaling system despite the continued presence of a stimulus. Phosphorylation of the GPCR by a serine or threonine kinase allows the arrestin to bind to the GPCR and become desensitized (Aubry & Klein, 2013).

G Proteins are the main proteins that GPCRs target. They are a subset of the signaling proteins from the Guanosine Nucleotide-Binding Proteins family. They belong to a larger group of enzymes called GTPases. The two major

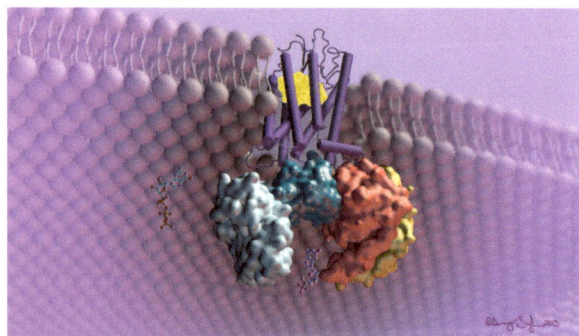

Figure 9: GDP to GTP exchange in a GPCR

IV DYNAMIC HELICES:

GPCRs respond to ligands by undergoing dynamic conformational changes in the transmembrane domain which alter the ability of the GPCR to communicate with intracellular signaling partners. These conformational changes in the transmembrane domain are shifts in position and orientation of certain helices within the helical bundle. The inactive state is stabilized by the close alignment and interaction of TM 3 and 6 at the intracellular face, a closed conformation which prevents interaction with signaling partner proteins. The overall movement of the GPCR can be visualized with a clothespin analogy. The agonist ligand acts as the pressure exerted on the top of the clothespin compressing it and allowing the bottom to open as the GPCR opens on its intracellular face (van der Kant & Vriend, 2014). Upon activation TM 6 swings and twists outwards away from TM 3; TM 5 moves in coordination with TM 6, and TM 7 moves and twists inward (Venkatakrishnan et al., 2014) (Figure 10). This opens up a crevice on the intracellular face to permit the binding of the intracellular signaling partners such as G Proteins and Arrestins. Subsequent intracellular signaling events and regulatory events such as phosphorylation and desensitization are initiated within or near the intracellular domain as a consequence of these helical shifts (Whalen, Rajagopal, & Lefkowitz, 2011).

the enzyme, other aspects of regulation were localized to distinct secondary sites of action, topographically removed from the active site. The information was relayed by conformational changes of the protein elicited by occupancy or modification of that secondary, i.e. "other" or "allosteric," site. The concept could be expanded to other non-enzymatic regulatory proteins as well, as long as two or more topographically distinct sites contributed in an interactive way to influence the function of that protein.

The first way that allostery can be understood to relate to GPCR biology is to define the relationship between ligand occupancy of the GPCR and the engagement of signaling partners at the intracellular binding surface of the GPCR as an allosteric relationship. We will use the term "allosteric network" to define this relationship (Wootten, Christopoulos, & Sexton, 2013). Ligands bind to specific binding sites, generally located either within the outer half of the transmembrane bundle or within the extracellular domain, and transmit information through a network of conformational rearrangements to the "topographically distinct" activation site, the intracellular surface of the receptor where binding to signaling partners will occur.

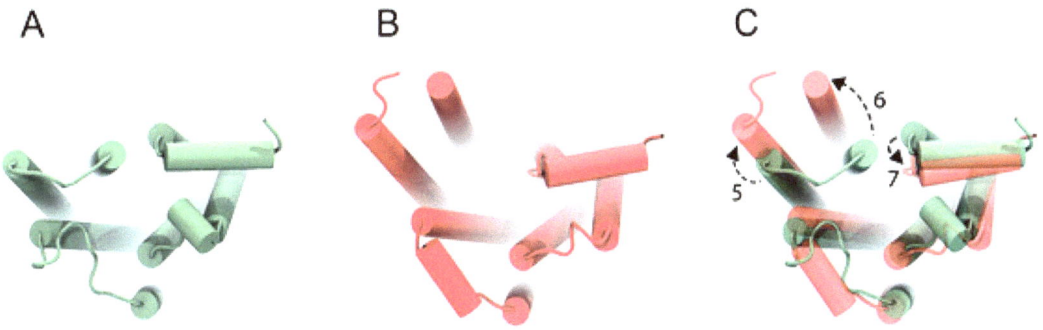

Figure 10: A. An inactive Beta 2 Adrenergic (PDB ID 4GBR). B. An active Beta 2 Adrenergic (PDB ID 3SN6). C. The dynamic rearrangements of TM 5,6 and 7 in the active vs inactive state of a GPCR. Image courtesy of Evelyn Maizels.

V ALLOSTERIC AND ORTHOSTERIC

The concept of allostery (also allosterism, adj allosteric) is key to the understanding of GPCR function. Several different aspects of GPCR biology can be described by the term "allostery," and we will consider each of these aspects separately and clarify how each these aspects contribute to the unique character of GPCR signaling (Smith & Milligan, 2010).

We must first define what is meant by allosteric. "Allo-" denotes "other" and its most general definition is: an interaction between two or more topographically distinct sites. These allosteric relationships were originally defined for enzyme activity: while some aspects of enzyme regulation could be accounted for by modifications at the active site of

GPCRs can also form ternary complexes with additional ligands or accessory protein and display altered binding and signaling properties (Wootten et al., 2013).

The second way allostery can relate to GPCR biology is through the concept that GPCRs may interact with different types of ligands in different ways. Rather than a single ligand (or type of ligand), being able to act upon GPCR to elicit changes through occupancy of a single unique ligand binding site, some ligands may act upon GPCRs by binding at other additional secondary "allosteric" sites.

An additional term "orthosteric," is defined and related to "allosteric." Novel ligands for GPCRs act at both orthosteric and allosteric sites to regulate the receptor's function. The orthosteric site is the site in which the major endogenous ligand binds to the GPCR. Ligands that act through the orthosteric site are classed as orthosteric ligands. Certain additional ligands may bind at the same orthosteric site as well. Ligands that bind to secondary sites are topographically distinct from the orthosteric site are classed as allosteric sites (Luttrell, 2014).

Ligands that act by binding to those secondary sites are "allosteric ligands." Binding at an allosteric site causes a change in the conformation, (i.e. the tertiary and quaternary structure) of the target protein, altering the interaction of that target protein with its binding partners, including orthosteric ligands and or intracellular signaling proteins, at sites which are distinct from that allosteric site. The additional allosteric sites provide receptor subtype selectivity in order to fine-tune the receptor signaling strength (Smith & Milligan, 2010) (Figure 11).

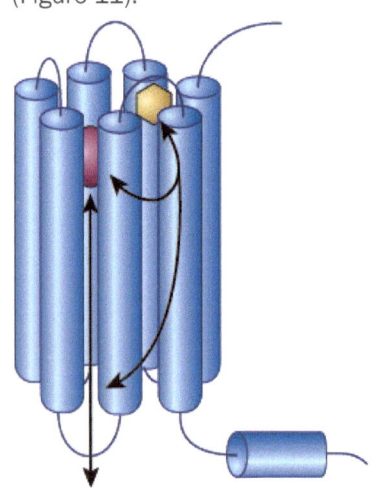

Figure 11: The difference between an Orthosteric ligand binding site (red) and an Allosteric ligand binding site (yellow), detailing how an allosteric modulator can alter the potency of an orthosteric ligand.

Allosteric ligands can be further classed into allosteric agonists or antagonists and allosteric modulators. An allosteric agonist or antagonist directly alters activation response of the receptor whether or not the orthosteric site is occupied. This causes a conformational change that alters the interaction of the receptor with its intracellular signaling proteins. In contrast, an allosteric modulator alters the interactions of the GPCR with its orthosteric ligand. An allosteric modulator has no effect on the GPCRs actions when the orthosteric site is empty. It only modulates the receptor response to the orthosteric ligand when the orthosteric site is occupied. In that way an allosteric modulator refines the response of the receptor to the presence of the orthosteric ligand.

Allosteric ligands can provide substantially greater GPCR subtype selectivity, while allosteric modulators can have a theoretical maximum effect on the orthosteric ligand function. This means that accessible ligand binding sites and cell surface location make GPCRs druggable, and allosteric sites expand the number of way in which GPCRs are druggable (Wootten et al., 2013). The conditions in which the effects of orthosteric ligands are used are expected to be intolerable because the receptor subtypes are widely expressed in GPCRs. This means that allosteric ligands mediate a wide and complex range of functions, and that the therapeutic windows of orthosteric ligands may be low, so it is more appropriate for pharmaceutical companies to target allosteric ligands. The allosteric effect can depend on the orthosteric ligand which makes predicting how allosteric ligands will behave extremely difficult (Smith & Milligan, 2010).

VI GPCRS FORMING DIMERS AND OLIGOMERS

The signal transduction protein receptors GPCRs are often described as monomers, but they can interact and form dimers and oligomers. Monomers are proteins that work as a single unit. This means that they work as a single chain without subunits, rather than as a part of a complex. Subunits can be referred to as protomers. Protomers are subunits of a larger complex. The term monomer can be used interchangeably with the term protomer in the context of a larger complex. One difference between the two terms is that a protomer can be a unit of an oligomeric protein which means that it can be defined as a single subunit or as several different subunits that form a larger whole. GPCRs are typically thought of as monomers, but they have been known to form dimers which are a macromolecular complex formed by two, usually non-covalently bound, macromolecules. GPCRs are also known to form larger oligomers, which are macromolecules that consist of a multitude of protomer subunits, again usually formed by non-covalent bonds (Ferre et al., 2014).

These oligomers can either be permanently stable, or work together once and disassociate. The consequences of complex formation for signaling are still an area of active research. Some evidence suggests that dimeric complexes have a reduced function as compared to monomers. However, other evidence suggests the opposite is true, that dimeric complexes have enhanced signaling potential (Milligan, 2013). Recent papers suggest a stoichiometry of two receptor protomers with one G Protein. The dimers within this complex may have asymmetric function, and one protomer out of the dimer may directly activate the G Protein while the other does not (Jacobson & Costanzi, 2012). Similarly a two receptor single arrestin complex has been described (Gurevich & Gurevich, 2013).

Two specific examples of interaction suggest that dimer formation is physiological. The first case shows that dimer pairs can complement receptor function. In defective mutants monomeric partners can act together as in the case of the Luteinizing Hormone (LH) receptor. A mutant LH receptor monomer that was signal-deficient was able to partner with a different mutant LH receptor monomer that could not bind to exogenous ligand, and complement each other's defects. This forms oligomers that function effectively. This is despite the fact that as single entities they cannot function (Rivero-Muller et al., 2010).

Dimers can also alter the signaling outcome as in the case of the dopamine receptor. The second case describes dopamine receptors D1 and D2 which are monomers that work that signal through Gαs and Gαi respectively. However the D1/D2, dimer changes the signal to target to Gαq. This is coupled with the fact that dopamine signaling in the brain is through Gaq meaning that the D1/D2 dimer is most likely found in vivo. This shows that the D1/D2 have novel function when co-activated in the same cell and it is indicative of the mechanism of the functional link observed between these two receptors in the brain (Lee et al., 2004). The increased level of calcium/calmodulin dependent protein kinases II in the nucleus indicates that the D1/D2 receptor dimer is actively signaling in this context, and thereby contributes to synaptic plasticity (Gurevich & Gurevich, 2013).

The signaling interfaces are net yet definitive or characterized. It is postulated that helices 1 and 8 form the dimer (Figure 12), and others that suggest helices 4 and 5. There is also the possibility that the monomers work as one long chain coupling across both interfaces to function as a larger oligomer (Ferre et al., 2014).

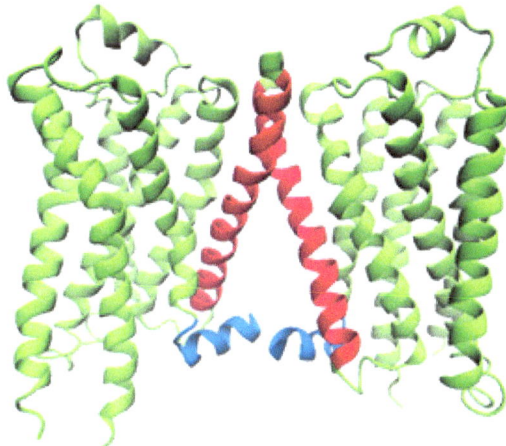

Figure 12: A Turkey Beta1-Adrenergic Receptor Dimer (PDB ID 4GPO asymmetric unit and assumed biological unit) with helix 1 (magenta) and helix 8 (blue) interaction interface

THE GUIDE FOR BEST PRACTICES: THE ACCURATE DEPICTION OF GPCRS

Having increased knowledge of GPCR structure, function, and conformational flexibility will aid in generating accurate images of GPCRs. The next portion of this guidebook will focus on applying GPCR knowledge to creating beautiful and didactic images. This section will discuss the use of specific viewpoints and styles of representation. It will also discuss commonly made errors and problems that an illustrator might run into when working with GPCRs.

I WHEN TO USE 2D OR 3D STYLES

There many different uses for illustrations of GPCRs and perhaps the most important initial distinction to make when illustrating GPCRs is whether to use a 2D representational image, a 3D model, or a hybrid of both. There is a very fine line between an oversimplified illustration and an overly complicated illustration, this can lead to illustrations that do not communicate the information necessary to instigate scientific discovery. Communicating information effectively is the most important objective when illustrating GPCRs. When beginning a GPCR illustration it is best to identify the audience and the purpose of the illustration. The audience and purpose for the illustration directly inform the decisions for which representation is most appropriate. Generally speaking stylized 2D illustrations can show complicated information about downstream signaling cascades more effectively than more realistic 3D structures which can obscure the information. 3D models are more effective for showing the interactions between the helices of the GPCRs and the ligands, and binding proteins. An example of when one might want to use a hybrid of 2D and 3D illustration techniques would be to show the helices of the GPCR, but still keep the downstream signaling information clear and informative. 2D and 3D considerations will be discussed later in this paper.

II PURPOSE AND AUDIENCE

There are three primary audiences this guidebook is targeting are research scientists, clinicians, and students. Research scientists have a significant background understanding of GPCRs, making it very important not to oversimplify these illustrations. Clinicians have an interest in how GPCRs work in terms of pharmacology; they also may be required to explain to patients how GPCRs work in terms of their medical requirements. Students are either starting to learn about GPCRs or have only a basic understanding of GPCR structure and function. In any of these cases, the purpose of illustrating a GPCR goes back to educating the audience. Scientific researchers would want to communicate their information to other researchers as well as educate them on the work that they are doing. Clinicians need to be educated on the effects of ligand binding and its effects on downstream signaling cascades. Students also need to be educated on the significance of GPCR structure, function, and conformational flexibility.

III DESIGNING THE STAGE

The canonical view of a GPCR is a transmembrane view through the membrane with helix 8 pointing to the right side of the image. This guidebook will define this as the standard view. There are many reasons why one would chose to illustrate a GPCR from a different view. The following sections will address the common representations standard, extracellular, and intracellular views and what each view is best suited to depict.

Standard: The standard view is used to show the big picture of GPCRs. This is due to a GPCR having functioning extracellular and intracellular portions it frequently happens that the best way to view a GPCR contextually is from the standard view. That way you can see how the ligand interacts with the GPCR as well as how the GPCR initiates downstream signaling. While helix 8 typically points to the right of the image with helices 5, 6, and 7 in the front of the GPCR there are instances when rotating the GPCR would be beneficial and enhanced the representations. An example of this would be to show how the G Alpha protein domains move when activation of the GPCR caused a docked G-protein alpha subunit to release its pre-bound guanine nucleotide GDP. This is best viewed when the GPCR is rotated 90 degrees toward the right from the standard view, bringing helices 3 and 4 to the front of the field of view.

Extracellular: The extracellular view is effective in showing where a ligand would bind within the GPCR. This is an important viewpoint when determining if a ligand is binding to the orthosteric site or at an allosteric site. Depending on where the ligand binds to the GPCR it can have different effects on the downstream signaling cascades. For example a classical ligand which binds to the major ligand binding pocket located between helices 3, 5, and 6 to activate a G-protein, while a biased ligand which binds to the minor binding pocket, contacting helices 2,3, and 7 is likely to activate arrestins without activating G proteins . An illustrator could be called on by a pharmaceutical company to show how a ligand from a new drug therapy would bind to the GPCR more effectively then the endogenous ligand.

Intracellular: The intracellular or, cytoplasmic, view is frequently used to show the movement of the GPCR helices. There are a lot of conformational differences between the helices in the active and inactive conformations of a GPCR. These changes are best viewed from the inferior position . For example, the ionic lock, a feature which may be present in the inactive state, can be clearly seen from the inferior side where it would be obscured from the standard or extracellular view.

IV STYLISTIC CHOICES

One of the biggest stumbling blocks when it comes illustrating GPCRs is choosing how to illustrate them. Is an schematic, a stylized, a 3D representation, or a 2D/3D hybrid more appropriate (Figure 13)? The answer to this question depends on what information is to be communicated. In many cases the iconic style of illustration would be appropriate, and there are also specialized diagrams that could be used. One example of this would be a snake plot which is used to visualize specific amino acids in the context of their helices. The second specialized diagram would be a top-view box plot which is used to visualize the top 20 amino acids in the GPCR.

Schematic Style: These are simple shapes to clarify a complex concept such as the downstream signaling cascades that GPCRs initiate. These can be better communication tools than a complicated fully rendered illustration. The schematic style is deceptively simple, but the icons are very important for showing difficult to understand chemical reactions and interactions with ligands and proteins. There are two styles of schematic illustration. One views the GPCR bundle of rods as a rectangle, or a filleted rectangle. The other one displays seven parallel bars as helices in homage to the iconic snake plot representation denoted in this paper as a snake icon.

Stylized: The stylized GPCR illustration is effective for showing how the helices interact with one another. Another use is to show how the ligand causes changes to the shape of the receptor.

3D Representation: The 3D representation model is frequently used to show a highly accurate representations of the GPCR. There are many different styles of 3D representation, and each one can be used in different types of illustrations (Figure 14). The Cartoon and NewCartoon models or Ribbon models are backbone representations that can be used to show the helical interactions clearly. They can impart specific tertiary or quaternary structural overviews while still appreciating contributions of secondary structure to the overall fold of the protein. They can be used to show the ligand binding placement to be understood in context. The Surface model can be used in animations to support the placement of a GPCR in a cellular environment or to create an atmospheric presentation that is appropriate to an editorial rather than didactic illustration. The surface model is also effective for indicating hydorphobicity and other important biochemical functions.

2D/3D Hybrids: 2D/3D Hybrids are effective, but they can be very tricky. The 2D/3D hybrid can show a 3D GPCR model and then simplifying the other components in 2D form. This is effective for truly highlighting the importance of the GPCR while still communicating the information.

LPA$_2$ and the role it plays in colorectal cancer

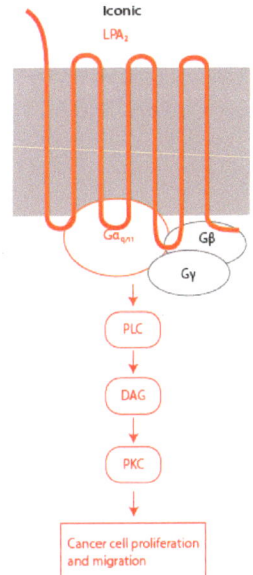
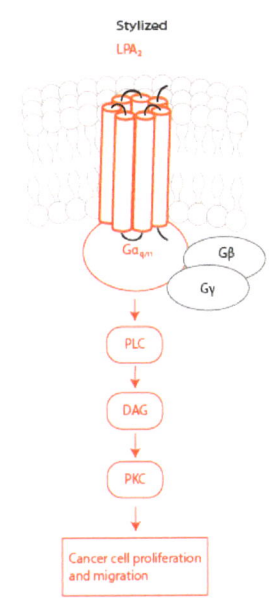
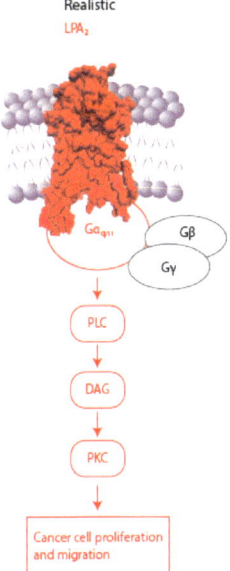

Figure 13: Iconic (Left), Stylized (Middle) and Realistic (Right) Representations for illustrating a GPCR

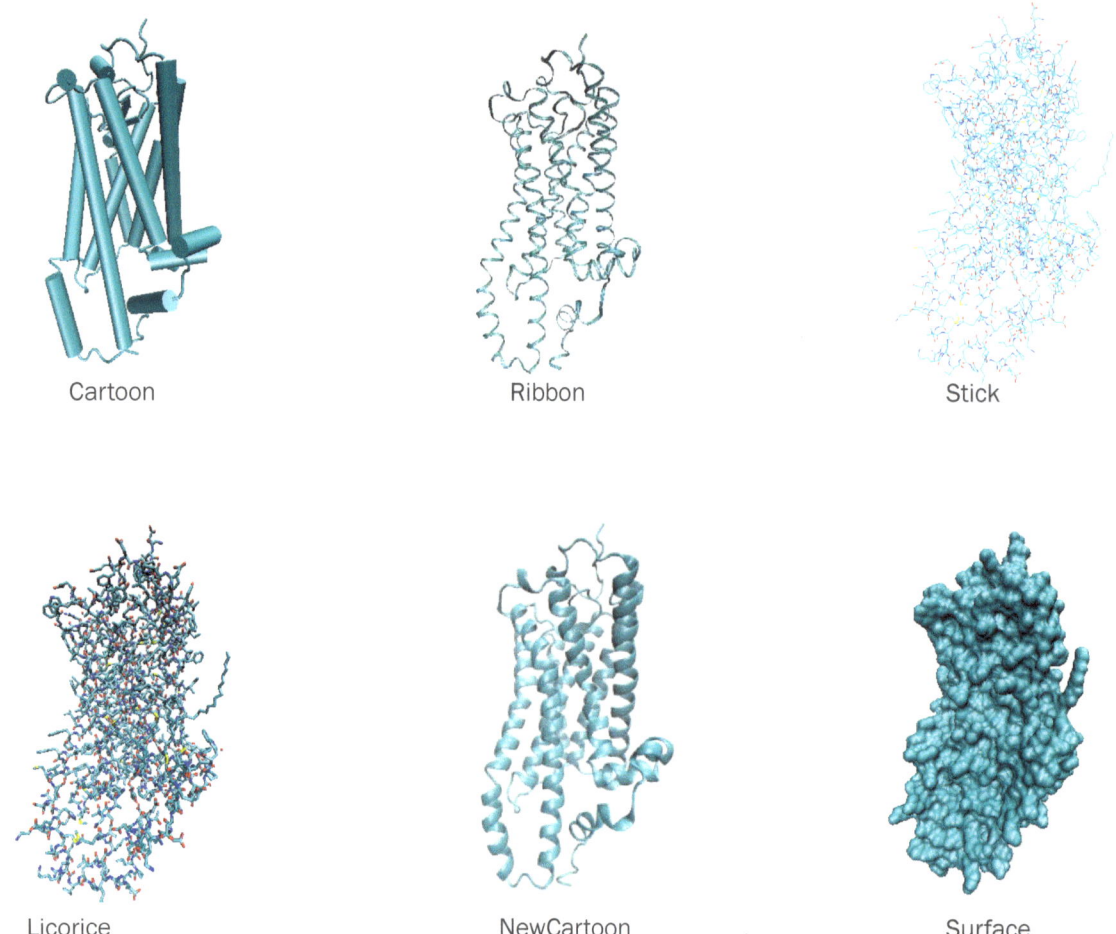

Figure 14: Different styles of 3D model representations available in the VMD.

V CHALLENGES AND PITFALLS

There are several stumbling blocks when illustrating GPCRs. Common errors include incorrectly represented 2D/3D hybrids, improperly arranged accordion models, and mixed or flipped orientation of GPCRs to the membrane and its own helices. These errors ensure confusion while studying GPCRs. They misrepresent how ligands bind to GPCRs, how the GPCR helices interact, and how the GPCRs are situated in the cell membrane. It is important to not make any of the common errors when illustrating GPCRs because it undermines their scientific integrity.

2D/3D Hybrids: The 2D/3D hybrids can be done correctly and incorrectly. A correct example of a 2D/3D hybrid would when using a 3D model of a GPCR in a 2D environment. An incorrect example of a 2D/3D hybrid would be showing a 3D environment but an iconic GPCR. This style confuses information too easily. Another common error which is similar to a 2D/3D hybrid is the 3D snake plot. This is when a GPCR 3D model is made, but then it is kept in the form of the snake plot, so it is lined up in the membrane. This is incorrect because in a 3D environment the GPCR should be in its tertiary protein structure. It forms a bundle of rods in order for ligands to bind to it, and for signaling proteins to bind to as well. This style of illustration fails to show the relationship of the helices to each other and the formation of the ligand binding pocket (Figure 15).

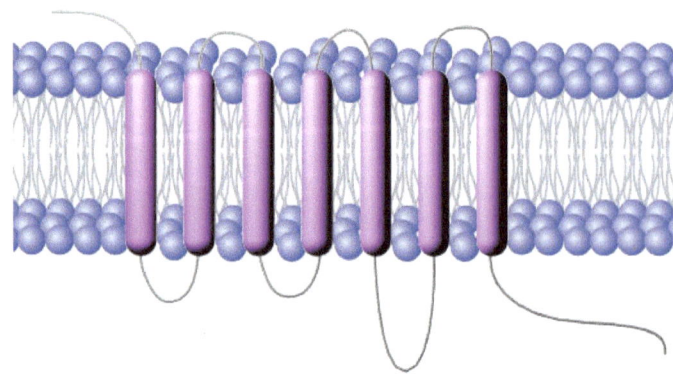

Figure 15: Common example of an incorrect 2D/3D hybrid.

Accordion Style: The accordion style of illustration is a diagrammatic representation of GPCRS that misrepresents their helical N-terminus to C-terminus arrangement. They repreresent an incorrect portrayal of the spatial relationship of the GPCR helices (Figure 16).

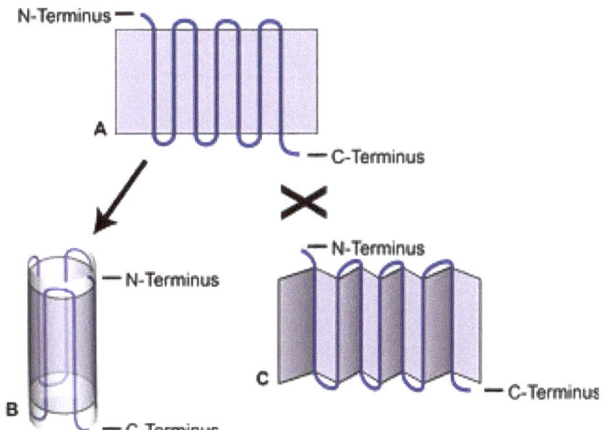

Figure 16: Visualization difficulties arise from associating the 2D iconic depiction with 3D spatial relationships. A shows the 2D snake icon. B shows the correct translation of the 2D iconic image incorporating 3D spatial relationships. C shows the incorrect accordion depiction which attempts to incorporate spatial information without accurately depicting the correct helical positioning.

In the accordion style of representation there are 2 rows of helices with every even-numbered helix presented as forward in the front row, and every intervening odd-numbered helix presented as behind, in the second row (Figure 17). This misrepresents the 3D conformation of the GPCR as well as the relation of the helices to one another. Additional variations of the basic accordion error, with alternative stacking orders assigned to the helices, also should be avoided for the same reasons.

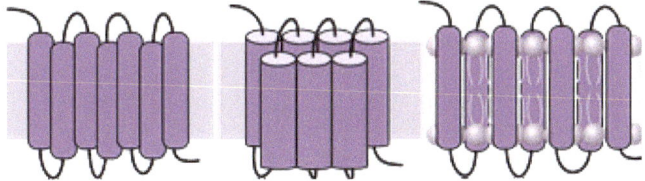

Figure 17: Three examples of incorrectly illustrated accordion style GPCRs.

Orientation: GPCRs are transmembrane proteins, so their position in the membrane is very important. As described in the Biochemistry section of this paper the individual GPCR helices span the phospholipid bilayer with the exception of helices 5 and 6 which have intracellular extensions that span beyond the membrane. Often GPCRs are depicted too big for the membrane, with all of the helices are extending outside of the membrane. This is incorrect because it does not show the true nature of the GPCRs, and can be confusing to audience members who are not familiar with the subject matter.

VI 2D BEST PRACTICES

2D illustration types include both familiar diagrams such as those used to present overviews for signaling pathways, as well as several types of specialized representations. For example the snake plot is a specialized 2D style of representation presenting aspects of both primary and secondary protein structure. The snake plot divides the GPCR into its helices and intracellular and extracellular loops by naming their specific amino acids (Figure 19A). A 2D snake plot is particularly useful to depict how specific amino acids are distributed within in the context of each particular helix. 2D snake plots demonstrate the order of specific amino acids that reside within particular helices to form conserved motifs controlling the activation state for GPCRs. There are several databases that can aid in the accurate depiction of 2D GPCR particularly the G Protein Coupled Receptor Database (GPCRDB) which allows users to download diagrams such as the snake plot for each GPCR.

Another specialized type of 2D presentation is the box plot; the box plot shows the first 20 amino acids in each helix from a superior or extracellular view. 2D box plots simplify the depiction of how helices relate at the extracellular face, and thereby can aid in understanding how ligand binding cavities are arranged among specific helices, and how specific amino acids near the extracellular interface may contribute to control the conformation of the GPCR. The GPCRDB provides users a method to generate box plots of GPCRs of interest as well (Figure 19, B).

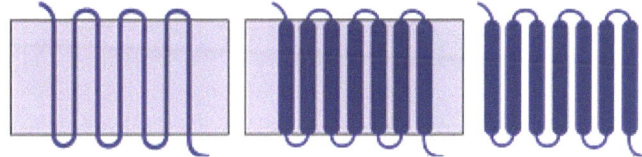

Figure 18: Three common forms of the snake icon.

When illustrating 2D GPCRs there are several different stylistic options. These options include the iconic and stylized versions of GPCRs. When choosing which style to use the audience of the illustration as well as the purpose of the illustration should be taken into consideration. The iconic illustrations can either be a rectangle or a filleted rectangle or a variation on the snake plot which we will denote as a snake icon. The snake icon is read from left to right or N-terminus to C-terminus (Figure 18). This can be further described using a sewing analogy. If you consider the knot at the end of the thread as the n terminus, the needle as the c terminus, and the fabric to be the cell membrane then the stitches would correlate to the GPCR. These diagrams can be created in Adobe Illustrator or Photoshop.

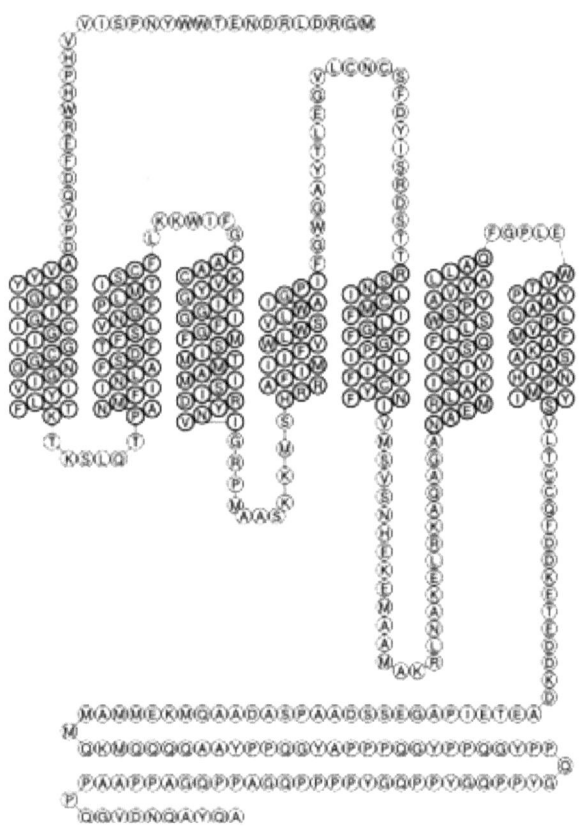
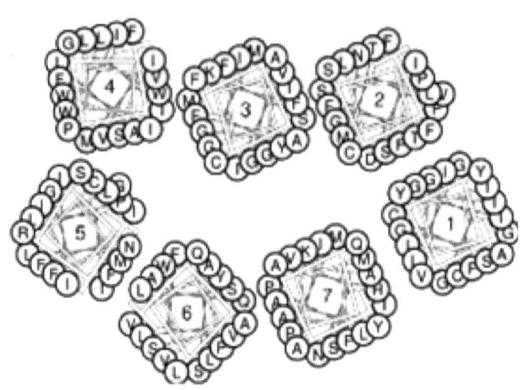

Figure 19A: Snake Plot displaying all amino acids in order indicating where they sit in the transmembrane helix or the intracellular or extracellular loops of Squid Rhodopsin (PDB ID 2Z73). 19B: The Box Plot showing the top view of the helixes and the first twenty amino acids in the GPCR Squid Rhodopsin (PDB ID 2Z73).

The stylized illustrations are similar to the 3D models in that they provide information on the helical representations of GPCRs. The stylized versions of GPCRs have formed the bundle of rods and are set up in their cylindrical shape. They don't have to show correct helical conformation because they are stylized. But it would be appropriate to illustrate them in the standard position with the N-terminus of the right of the illustration with helix 1 going behind and forming the cylinder with helices 5, 6, and 7 in the front of the illustration.

VII DATABASES

There are a variety of databases that contain information on GPCRs. This paper will focus on four that, when used in conjunction with each other are useful in the visualization of GPCRs in 2D and 3D. The four resources include The Universal Protein Resource (UniProt), The G Protein Coupled Receptor Data Base (GPCRDB), The Orientations of Proteins in Membranes Data Base (OPM) and The RCSB Protein Data Bank (PDB). A description of each database and its key features for visualizing a human Beta-2 adrenergic GPCR will follow.

A. The UniProt Knowledge Base (UniProtKB)

The UniProt Knowledge Base (UniProtKB) (http://www.uniprot.org) (UniProt Consortium, 2014) has mapped out the known protein primary sequences. The primary sequence is the list of amino acids that comprise a protein. While the primary sequence itself is not very predictive of the final tertiary or quaternary structure, it can serve as a building block for further visualizations. The primary sequence is represented as a string of UniProt sequence numbers from the N-terminus to C-terminus of the protein's polypeptide chain. UniProtKB also presents specific features of interest in a protein, such as its structural motifs, and known binding sites. It also starts to map out the secondary structure showing helices and turns. Users can also use the UniProt BLAST tool to determine similarities between multiple GPCRs. An important feature that UniProt provides is the assignment of a unique identification number to each polypeptide known as the UniProt accession code which allows users to look up the same GPCR in different databases. UniProt accession code for Beta-2 adrenergic receptor is P07550.

B. G Protein Coupled Receptor Database (GPCRDB)

The GPCRDB focuses solely on GPCRs, and features a comprehensive toolset in terms of research literature and visualization. One of the strengths of the GPCRDB (http://tools.gpcr.org/) (Isberg et al., 2014) is a set of tool that allows for various visualizations of the protein. Some of the more useful visualization tools are the diagrams of snake plots and box plots. These representations are useful for visually identifying the previously discussed amino acid conserved structural motifs including the D[E]RY, NPxxY, and CPxW. (see above, Section I The Science of GPCRs) The UniProt accession code can be used to access the information in the GPCRDB. The GPCRDB primary sequence diagram (Figure 20) maps out helices and intra- and extracellular loops, and correlates these structural features with the UniProt sequence numbers.

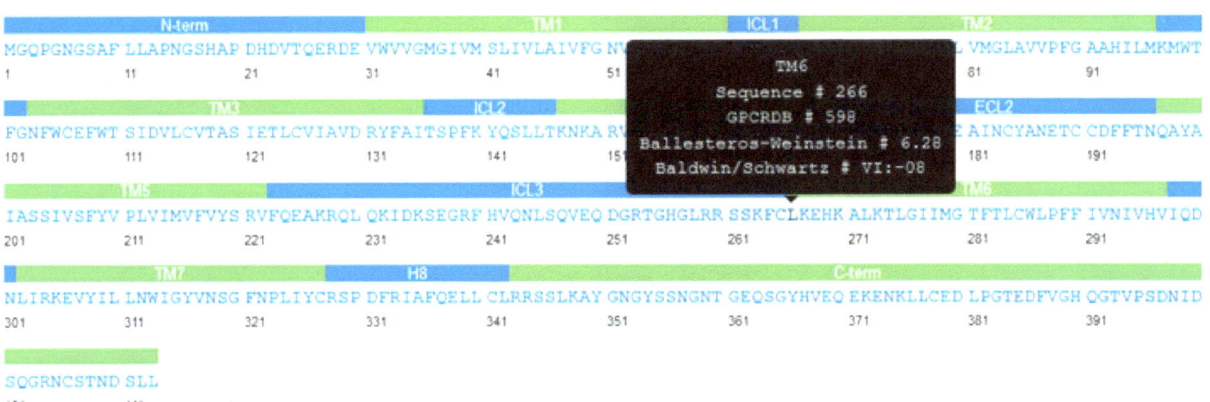

Figure 20: GPCRDB Primary Sequence diagram

C. Orientations of Proteins in Membranes

The Orientations of Proteins in Membranes (OPM) database (http://opm.phar.umich.edu/) (Lomize, Pogozheva, Joo, Mosberg, & Lomize, 2012) catalogs the orientation and position membrane proteins with respect to the phospholipid bilayer. The OPM can be used to accurately situate a GPCR within the cell membrane. The OPM can calculate and present the proper size relationships and orientation of the GPCR and associated proteins with respect to the hydrophobic core of phospholipid bilayer. The extracellular hydrophobic boundary is depicted in red, and the intracellular hydrophobic boundary is depicted in blue. Note that the provided markers denote the boundaries between the hydrophobic tail and hydrophilic headgroup regions of the associated phospholipid molecules, rather than denoting the outermost dimension of entire bilayer (Figure 21). This information is useful for accurately depicting a GPCR in its environment and accurately associating it with a phospholipid membrane bilayer in a 3D program.

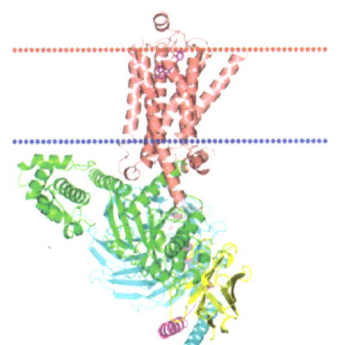

Figure 21: Diagram from the OPM showing the correct relationship of the GPCR within the context of the membrane (note: blue and red circle icons do not represent the size of the phospholipid heads, only a bounding box of the hydrophobic core of the membrane)

D. Protein Databank

The RSCB PDB (http://www.pdb.org) (Berman, Kleywegt, Nakamura, & Markley, 2013) has a wealth of information that can help visualize GPCRs. The RSCB PDB, as a member of the worldwide PDB, catalogs structural information for proteins whose structures have been experimentally solved by X-ray crystallography, electron microscopy, or NMR. The RSCB PDB website provides access to download PDB files to serve as a basis for visualization of the 3D structure for proteins of interest. The RSCB PDB website can also help the user to identify the important ligands that bind to GPCRs. Each PDB entry also contains a description of the molecules that are crystallized in the entry. The Molecular Description panel on the PDB entry page can help to identify any chimera proteins, which are sometimes used to stabilize GPCRs in order to crystalize them successfully. The accessory sequences are added to aid in the crystallization to make the chimera protein. The accessory proteins should not be included in the visualization of GPCRs unless they are relevant to the point of the illustration.

The visualization of components of the beta-adrenergic receptor-Gs complex contained within PDB file 3SN6 will be described below. A section of the abbreviated "protein feature view" displayed on the RSCB PDB page. This feature shows a particular chain within the complex, the component/PDB sequences are indicated by blue bars, and are mapped to show how they correspond to source sequences (UniProt accession code, followed by grey bar representing the UniProt sequence). In this molecule is Chain R contains the GPCR, so it is that chain being examined (Figure 22). Chain R is a chimera that is mapped out to two separate UniProt sequences. Chain R consists of two organisms which indicates that it is chimera. The top section of figure 25 depicts how some residues correspond to lysozyme P00720 (accessory sequence, included as crystallization aid), and the bottom section depicts how other residues correspond

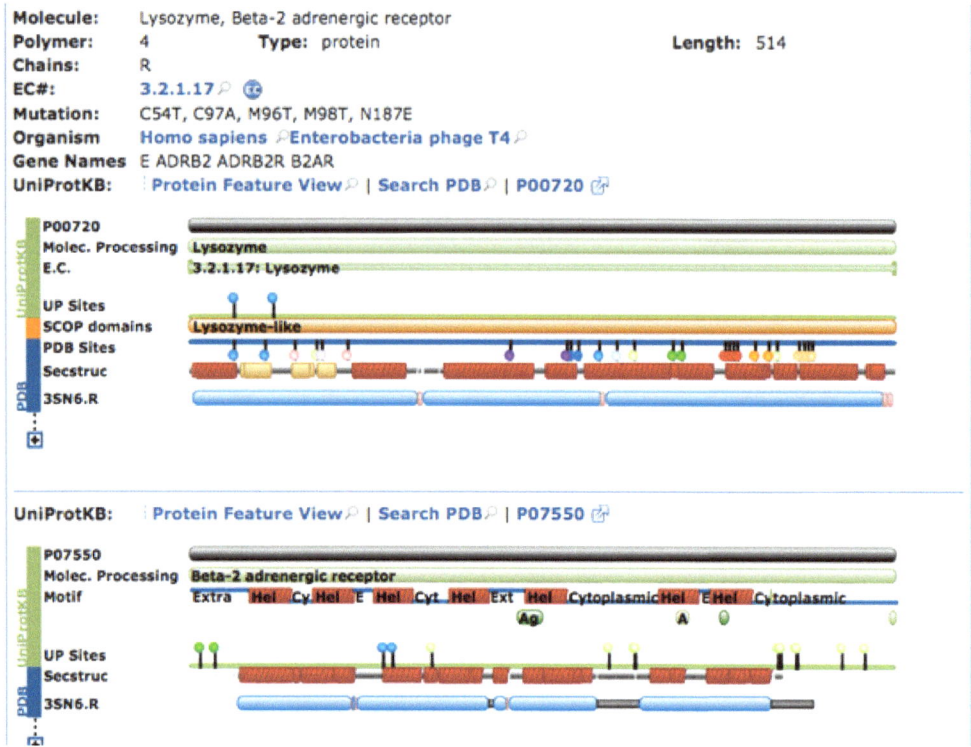

Figure 22: Protein feature view from Beta2-Adrenergic receptor (PDB ID 3SN6) showing the composition of Chain R

to beta2 adrenergic receptor P07550 (main sequence of interest).

Additionally the abbreviated "protein feature view" can be very useful to identify PDB files that represent only small fragments of the sequence of interest. Such fragments are encountered in the GPCR research arena, in cases where extended extracellular domains have been crystallized independently of the corresponding transmembrane regions of particular receptors. In such cases the blue PDB bar would cover only a small portion of the grey UniProt sequence bar.

E. PDB Header

The header portion of the PDB file (also available as a separate file for viewing or download) provides important details about what is in the PDB file. The PDB header file includes information on how to identify the features to help visualize the best GPCR. The header file provides the information on how to identify the features to help visualize the best GPCR. Such features include identities of the molecules contained in the structure file, specifying individual protein polypeptide chains as well as associated small molecules such as ligands, and providing sequence information to locate secondary structure features such as helices and beta strands, as well as modified or covalently bonded amino acids. The header file has a wealth of information that is pertinent to crystallizers, but not all of this information is necessary for illustrators to understand or use. The important portions for illustrators are COMPND and SOURCE records within the Title section, DBREF records with in the Primary Structure section, HET and HETNAM records within the Heterogen section, and SSBONDS records within the Connectivity Annotation section.

The COMPND and SOURCE records within the Title section provide details on the molecules and chains that are crystallized. A review of this section can confirm the presences of chimera proteins. Frequently the header file will label chimera, but it is possible to tell if a protein uses a chimera by the presence of two sequences from different sources in one molecule. This is illustrated for the beta2 adrenergic receptor example, as seen in the PDB header file for 3SN6 below. Chain R represents the GPCR component of the complex. The molecule ID for Chain R is 4, and there are two sequences from different species that are merged to constitute it. There is a main sequence of interest, which is the receptor, and an accessory sequence, which represents the crystallization aid. Now that the accessory sequence Lysozyme (T4 Phage) has been identified steps can be taken to isolate and remove it if desired (Figure 23).

Figure 23: The CMPND and SOURCE record of the PDB ID Header file showing Chain R of Mol_ID 4

The DBREF record is part of the Primary Structure section and has relevant information for trying to visualize the components of the protein including the UniProt sequence numbers (Figure 23: C), accession codes (Figure 23: B), and PDB resid numbers (Figure 23:A). The PDB resid number is the sequence number of the residues within a peptide chain as defined by the PDB. PDB resid numbers are assigned based on the primary structure, corresponding to the polypeptide amino acid sequence of the protein. In the VMD the PDB resid number can be used to make selections in order to manipulate the GPCR including determining the start and end points of a specific chain, helix, etc.

A B C

```
DBREF  3SN6 R   29   365  UNP    P07550   ADRB2_HUMAN    29    365
DBREF  3SN6 R 1002  1163  UNP    P00720   LYS_BPT4        2    161
```

Figure 23: The DBREF record of the PDB Header file showing A. the PDB resid numbers B. The accession codes C. UniProt sequence numbers

Frequently this PDB resid number is the same as the UniProt sequence number, but they are not always assigned the same value. For example chimera accessory sequences are typically offset from the UniProt sequence number by 1000. The reason that they are assigned offset numbers to readily identify them. In this way the DBREF section of the header files informs the user of the PDB resid numbers, which can be used in the VMD to exclude the chimera accessory sequence. In the Beta2-Adrenergic receptor 3SN6 example the two sequences that make up Chain R are derived from two sources: UniProt accession code P07550 (receptor; ADRB2_HUMAN) and accession code P00720 (lysozyme; LYS_BPT4). The beta2-adrenergeric receptor sequences correspond to PDB resid numbers 29 to 365 of Chain R. The lysozyme sequence corresponds to PDB resid numbers 1002 to 1163 of Chain R.

The Heterogen section includes the record types HET and HETNAM. These records define the non-standard residues including modified amino acids and chemicals including ligands. The Heterogen section is most useful for identifying and isolating ligands. The HET record assigns a chain identifier and PDB resid number to the heterogen that can be used to visualize it separately in the VMD. The HETNAM record provides more detailed information regarding the HET entry by specifically identifying the heterogen by chemical name. Figure 24 shows the Beta2-Adrenergic receptor file, the HET record shows that the heterogen that is the ligand activating the GPCR, abbreviated as POG, assigned to Chain R and PDB resid 366.

```
HET    POG  R 366      27
HETNAM     POG 8-[(1R)-2-{[1,1-DIMETHYL-2-(2-METHYLPHENYL)
HETNAM   2 POG  ETHYL]AMINO}-1-HYDROXYETHYL]-5-HYDROXY-2H-1,4-
HETNAM   3 POG  BENZOXAZIN-3(4H)-ONE
```

Figure 24: The heterogen section of the PDB Header file calling out the HET and HETNAM records

The Connectivity Annotation Section contains the SSBOND record type, which defines the disulfide cysteine bonds, and identifies the cysteine PBD resid numbers. This is important because GPCRs have a conserved cysteine on helix three, which helps to stabilize GPCRs. For the Beta2-Adrenergic receptor that is shown here (Figure 24), the cysteine bond that keeps helix three in place on GPCRs. One of two disulfide bonds in the R chain, is between the cysteine designated as Chain R PDB resid 106, and the cysteine designated as chain R PDB resid 191.

VIII 3D REPRESENTATIONS

There is a wealth of information and resources available to those who want to generate accurate 3D visuals of GPCRs. The information available is focused towards crystallographers and researchers. It may be difficult to decipher, but the resources available can be understood with guidance, knowledge and practice.

The freely available Visual Molecular Dynamics molecular visualization program was designed to be compatible with the databases discussed above. This section will discuss the preferred method for manipulating GPCR PDB files in the VMD.

Use of the PDB Header file and the GPCRDB sequence display at the same time provides an efficient way to visualize GPCRs. Cross referencing the information in both databases aids in the manipulation of the PDB entry with in the VMD.

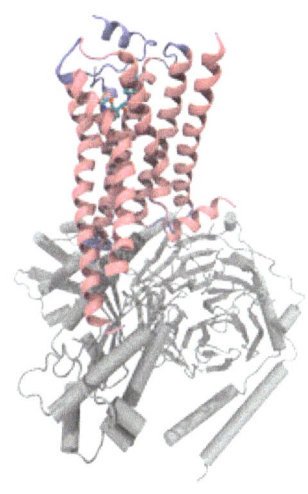

Figure 25: Final Beta2-Adrenergic receptor (PDB ID 3SN6)

The following will serve as a tutorial to isolating out the chimera protein and creating a specific representation. Figure 25 shows the final image created in the VMD. This image shows GPCR helices, loops, ligand, and heterotrimeric G protein subunits. This image will not be including any crystallization aids because the crystallization aids are not part of the signaling story that this image is telling.

Start by setting up the VMD scene with the graphic representation window open and have the PDB header file open to separate out specific polypeptide chains: Chain N is defined in the PDB Header files as a nanobody, which is a second accessory crystallization aid. Chain N is not wanted in this final image as it will not be showing any crystallization aids. Having now separated Chains A, B, and G, which represent the Alpha, Beta, and Gamma subunits of the Heterotrimeric G Protein and keeping the receptor Chain R on its own layer makes manipulation of the GPCR simpler (Figure 26).

Figure 26: Setting up the users work station with the VMD and PDB Header file

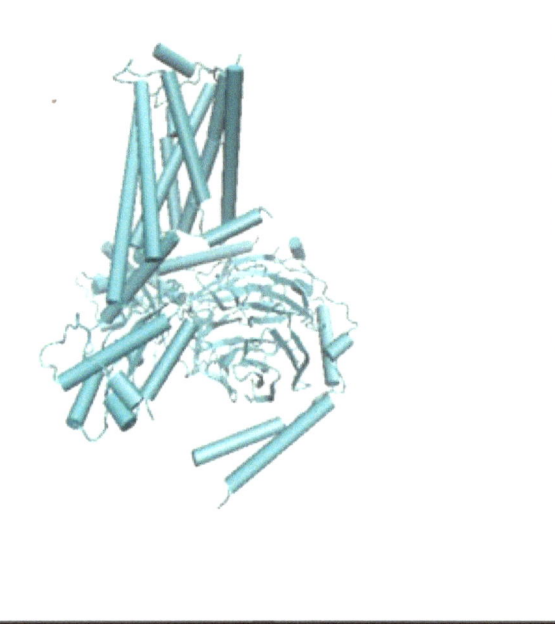

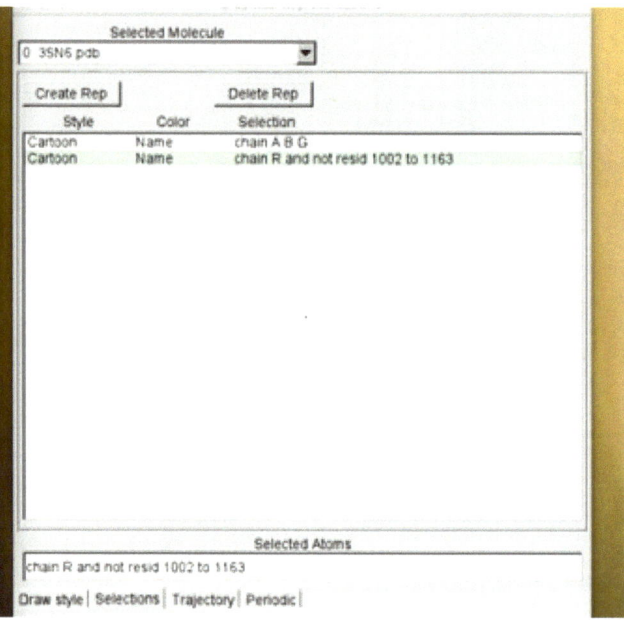

Figure 27: Removing the accessory sequence in Chain R in the VMD

The next step is to remove the accessory lysozyme sequence portion in Chain R. In the PBD header file under the DBREF record the lysozyme sequence corresponds to the PDB resid numbers 1002 to 1163. In the VMD the "not resid" selection method can be used to exclude the lysozyme sequence (Figure 27).

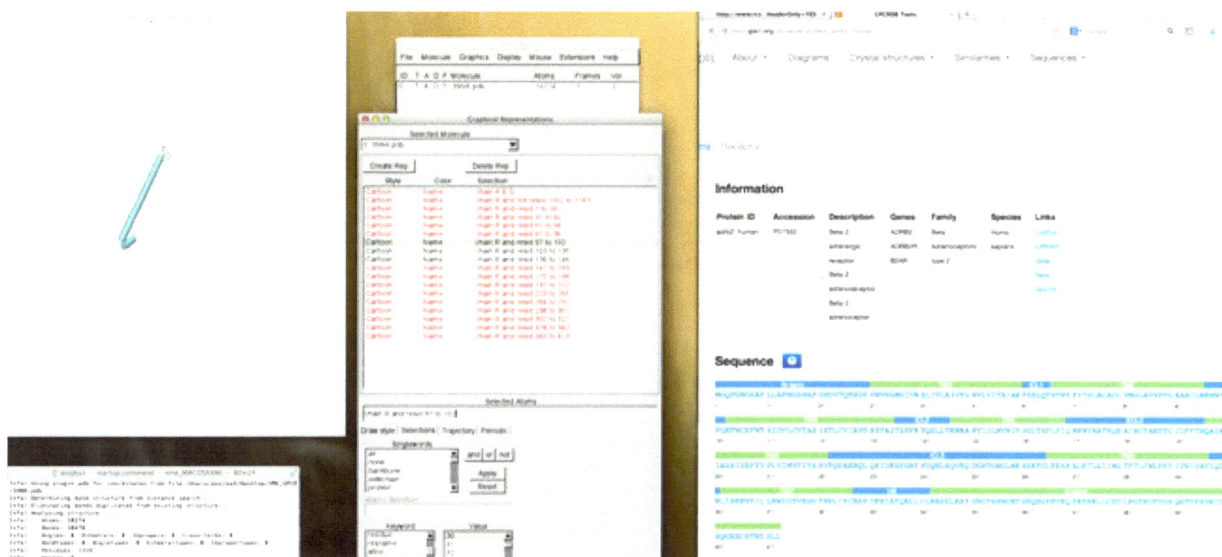

Figure 28: Separating out GPCR helices in the VMD to show an individual helix using the GPCRDB

The resid selection method is very useful in conjunction with the GPCRDB. Since the GPCRDB has the primary sequence of the Beta2-Adrenergic receptor each individual helix and/or loop can be identified. In this example helix three has been isolated, together with the contiguous extracellular loop ECL1 and the contiguous intracellular loop ICL2 sequences (Figure 28).

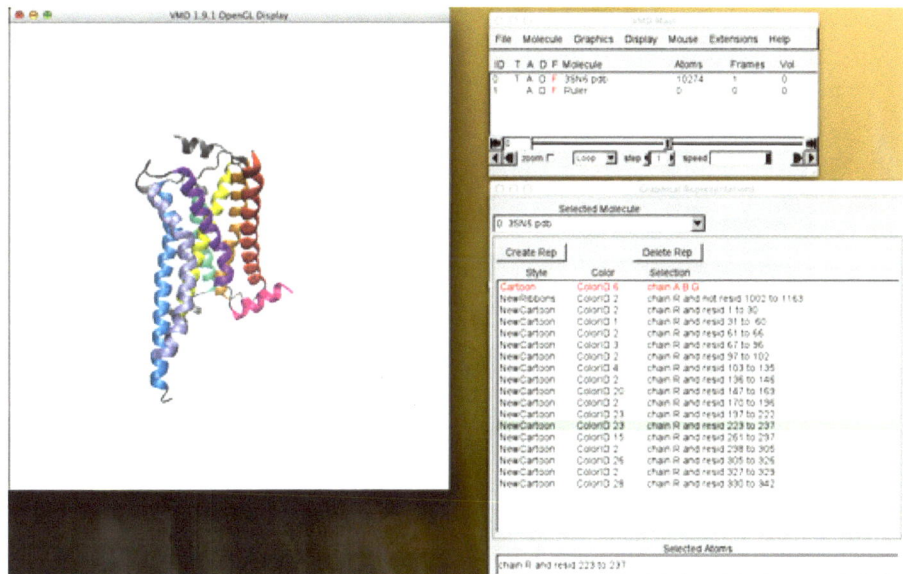

Figure 29: Showing each helix now separated and distinct from the others

A reason that it is beneficial to be able to parse out each individual helix to easily show how each individual helix interacts with the other helices (Figure 29).

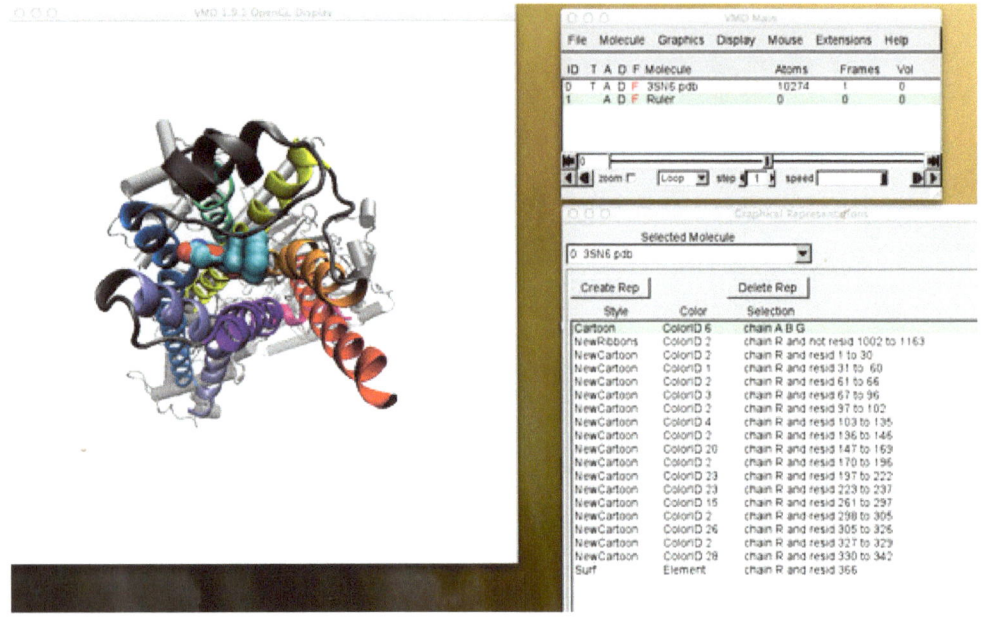

Figure 30: Showing the ligand in the GPCR binding pocket

The HET record of the PDB Header file provides information to designate the ligand using the selection command "chain R and resid 366." This parses out the ligand that is activating the GPCR as seen above (Figure 30).

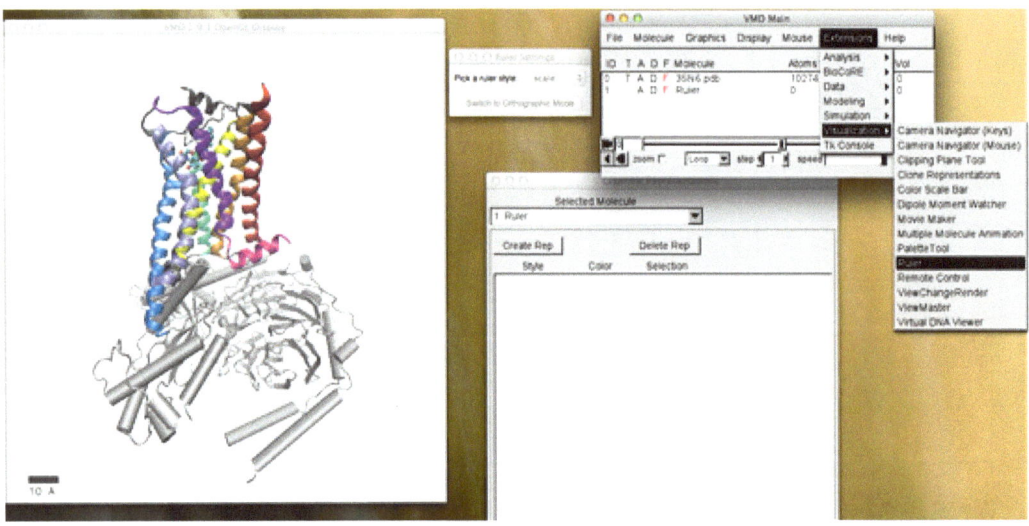

Figure 31: Using the ruler tool in the VMD to ensure the accurate size relations of the GPCR

In molecular illustrations size relationships often become drastically exaggerated and distorted. VMD offers a ruler tool to evaluate proper size relationships (Figure 31).

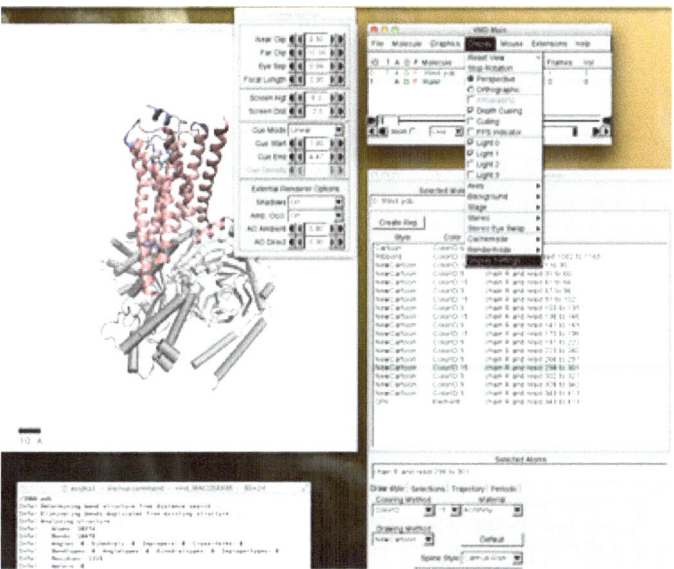

Figure 32: Using the Display Setting in the VMD to use ambient occlusion

The VMD allows for the generation of 2D images as well as 3D models. The VMD has several advanced rendering options in the Display Settings menu. Some of these options are depth of field and ambient occlusion. The incorporation of these rendering options in conjunction with the variety of rendering engines can allow for the creation of publication-quality images (Figure 32).

After creating all the needed representations you Graphics Representations window should appear as in (Figure 33).

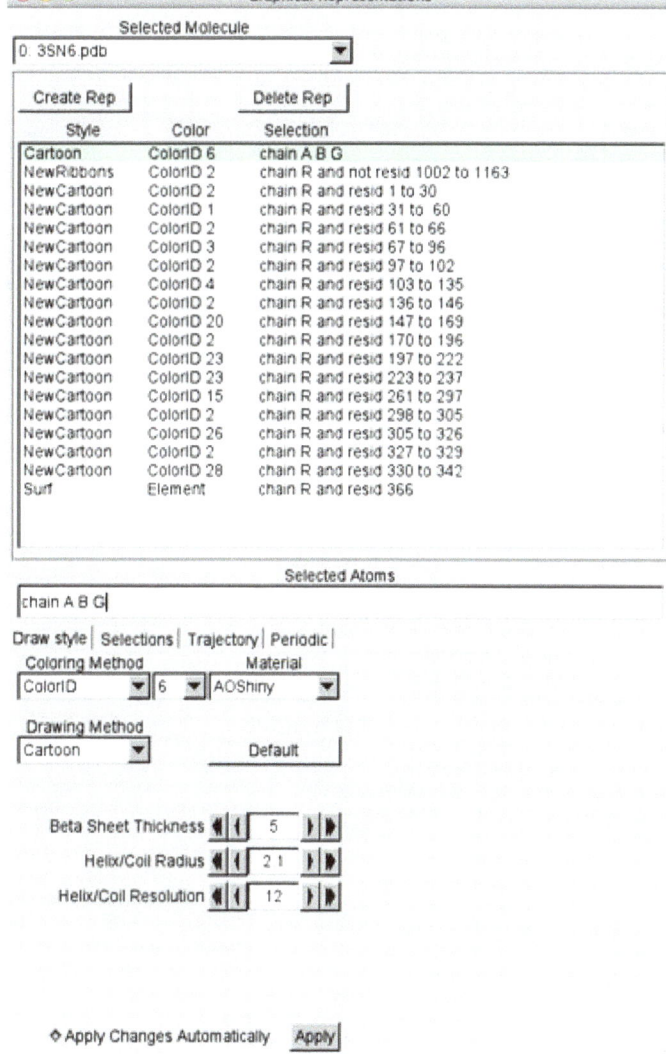

Figure 33: The final Graphics Representation Panel in the VMD showing the style and color of the selection.

The VMD has several rendering options. In order to render a still image one can use, among others, the POV-ray, Tachyon or Tachyon (internal render). In order to generate a model mesh for importing into 3DS Max one can export a Wavefront (OBJ and MTL) or STL (triangle mesh only) within the render dialog box. In order to export an OBJ file it is useful to reset the position of the molecule for each render via Display – Reset view. This ensures that the chains exported separately will maintain their position in relation to each other. VMD allows users to save visualization states (.vmd) to save selections and representations (Figure 34).

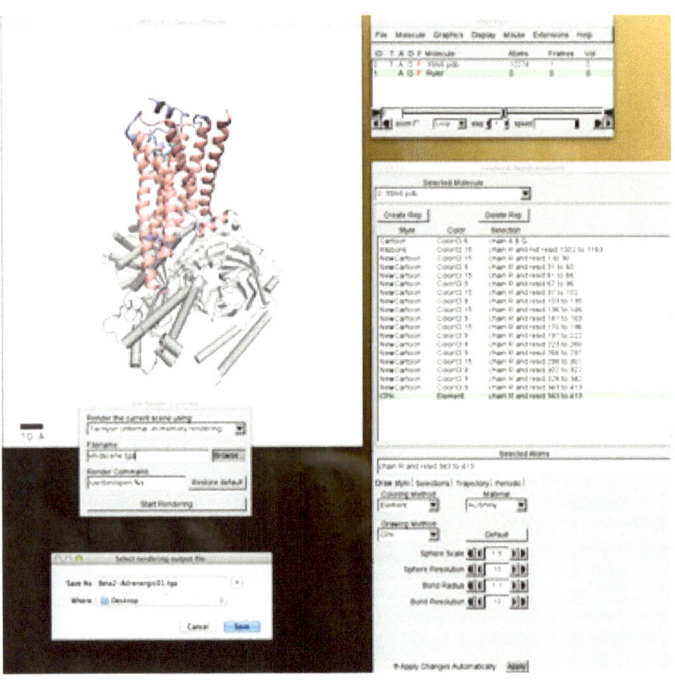

Figure 34: Rendering options in the VMD

CONCLUDING REMARKS

This research creates a knowledge resource for professional and student medical illustrators and guide to best practices when illustrating GPCRs. This information can be used to promote further discovery in the field of GPCR. By increasing the knowledge and fundamental understanding of a complex family of receptors that are continued targets of research and treatment. The GPCR Essentials and illustration guide support the integrity of medical illustrator's visual communication. The visual communication of complex concepts in biomedical research and drug development supports and strengthens the translation of medical information from bench to bedside and back to the bench to advance healthcare (National Institutes of Health. (n.d.)).

ACKNOWLEDGEMENTS

I would like to thank the faculty at UIC especially those members of my committee Christine Young MA, CMI, FAMI, Evelyn Maizels MD, PhD, MS, Kevin Brennan MS, and Leah Lebowicz MS.

WORKS CITED REFERENCES

Andresen, B. T. (2011). A pharmacological primer of biased agonism. Endocrine, Metabolic & Immune Disorders Drug Targets, 11(2), 92-98.

Aubry, L., & Klein, G. (2013). True arrestins and arrestin-fold proteins: A structure-based appraisal. Progress in Molecular Biology and Translational Science, 118, 21-56. doi:10.1016/B978-0-12-394440-5.00002-4; 10.1016/B978-0-12-394440-5.00002-4

Berman, H. M., Kleywegt, G. J., Nakamura, H., & Markley, J. L. (2013). How community has shaped the protein data bank. Structure (London, England : 1993), 21(9), 1485-1491. doi:10.1016/j.str.2013.07.010 [doi]

Caltabiano, G., Gonzalez, A., Cordomi, A., Campillo, M., & Pardo, L. (2013). The role of hydrophobic amino acids in the structure and function of the rhodopsin family of G protein-coupled receptors. Methods in Enzymology, 520, 99-115. doi:10.1016/B978-0-12-391861-1.00005-8 [doi]

Chini, B., & Parenti, M. (2009). G-protein-coupled receptors, cholesterol and palmitoylation: Facts about fats. Journal of Molecular Endocrinology, 42(5), 371-379. doi:10.1677/JME-08-0114 [doi]

DeWire, S. M., Ahn, S., Lefkowitz, R. J., & Shenoy, S. K. (2007). Beta-arrestins and cell signaling. Annual Review of Physiology, 69, 483-510. doi:10.1146/annurev.ph.69.013107.100021

DeWire, S. M., Yamashita, D. S., Rominger, D. H., Liu, G., Cowan, C. L., Graczyk, T. M., . . . Violin, J. D. (2013). A G protein-biased ligand at the mu-opioid receptor is potently analgesic with reduced gastrointestinal and respiratory dysfunction compared with morphine. The Journal of Pharmacology and Experimental Therapeutics, 344(3), 708-717. doi:10.1124/jpet.112.201616; 10.1124/jpet.112.201616

Ferre, S., Casado, V., Devi, L. A., Filizola, M., Jockers, R., Lohse, M. J., . . . Guitart, X. (2014). G protein-coupled receptor oligomerization revisited: Functional and pharmacological perspectives. Pharmacological Reviews, 66(2), 413-434. doi:10.1124/pr.113.008052; 10.1124/pr.113.008052

Gonzalez, A., Cordomi, A., Matsoukas, M., Zachmann, J., & Pardo, L. (2014). Modeling of g protein-coupled receptors using crystal structures: From monomers to signaling complexes. Advances in Experimental Medicine and Biology, 796, 15-33. doi:10.1007/978-94-007-7423-0_2

Goodman and Gilman's The Pharmacological Basis of Therapeutics, Twelfth Edition McGraw-Hill Professional;

12 edition (December 20, 2010) Editor: Laurence L. Brunton, PhD

Gurevich, V. V., & Gurevich, E. V. (2013). Structural determinants of arrestin functions. Progress in Molecular Biology and Translational Science, 118, 57-92. doi:10.1016/B978-0-12-394440-5.00003-6; 10.1016/B978-0-12-394440-5.00003-6

Heng, B. C., Aubel, D., & Fussenegger, M. (2013). An overview of the diverse roles of G-protein coupled receptors (GPCRs) in the pathophysiology of various human diseases. Biotechnology Advances, 31(8), 1676-1694. doi:10.1016/j.biotechadv.2013.08.017; 10.1016/j

Isberg, V., Vroling, B., van der Kant, R., Li, K., Vriend, G., & Gloriam, D. (2014). GPCRDB: An information system for G protein-coupled receptors. Nucleic Acids Research, 42 (Database issue), D422-5. doi:10.1093/nar/gkt1255; 10.1093/nar/gkt1255

Jacobson, K. A., & Costanzi, S. (2012). New insights for drug design from the X-ray crystallographic structures of G-protein-coupled receptors. Molecular Pharmacology, 82(3), 361-371. doi:10.1124/mol.112.079335; 10.1124/mol.112.079335

Katritch, V., Cherezov, V., & Stevens, R. C. (2013). Structure-function of the G protein-coupled receptor superfamily. Annual Review of Pharmacology and Toxicology, 53, 531-556. doi:10.1146/annurev-pharmtox-032112-135923; 10.1146/annurev-pharmtox-032112-135923

Kenakin, T. (2011). Functional selectivity and biased receptor signaling. The Journal of Pharmacology and Experimental Therapeutics, 336(2), 296-302. doi:10.1124/jpet.110.173948; 10.1124/jpet.110.173948

Lee, S. P., So, C. H., Rashid, A. J., Varghese, G., Cheng, R., Lanca, A. J., . . . George, S. R. (2004). Dopamine D1 and D2 receptor co-activation generates a novel phospholipase C-mediated calcium signal. The Journal of Biological Chemistry, 279(34), 35671-35678. doi:10.1074/jbc.M401923200 [doi]

Lomize, M. A., Pogozheva, I. D., Joo, H., Mosberg, H. I., & Lomize, A. L. (2012). OPM database and PPM web server: Resources for positioning of proteins in membranes. Nucleic Acids Research, 40(Database issue), D370-6. doi:10.1093/nar/gkr703; 10.1093/nar/gkr703

Luttrell, L. M. (2014). Minireview: More than just a hammer: Ligand 'bias' and pharmacological discovery. Molecular Endocrinology (Baltimore, Md.), , me20131314. doi:10.1210/me.2013-1314

Millar, R. P., & Newton, C. L. (2010). The year in G protein-coupled receptor research. Molecular Endocrinology (Baltimore, Md.), 24(1), 261-274. doi:10.1210/me.2009-0473; 10.1210/me.2009-0473

Milligan, G. (2013). The prevalence, maintenance, and relevance of G protein-coupled receptor oligomerization. Molecular Pharmacology, 84(1), 158-169. doi:10.1124/mol.113.084780; 10.1124/mol.113.084780

Moreira, I. S. (2014). Structural features of the G-protein/GPCR interactions. Biochimica Et Biophysica Acta, 1840(1), 16-33. doi:10.1016/j.bbagen.2013.08.027; 10.1016/j.bbagen.2013.08.027

Oldham, W. M., & Hamm, H. E. (2008). Heterotrimeric G protein activation by G-protein-coupled receptors. Nature Reviews. Molecular Cell Biology, 9(1), 60-71. doi:nrm2299 [pii]

Preininger, A. M., Meiler, J., & Hamm, H. E. (2013). Conformational flexibility and structural dynamics in GPCR-mediated G protein activation: A perspective. Journal of Molecular Biology, 425(13), 2288-2298. doi:10.1016/j.jmb.2013.04.011; 10.1016/j.jmb.2013.04.011

Rasmussen, S. G., DeVree, B. T., Zou, Y., Kruse, A. C., Chung, K. Y., Kobilka, T. S., . . . Kobilka, B. K. (2011). Crystal structure of the beta2 adrenergic receptor-gs protein complex. Nature, 477(7366), 549-555. doi:10.1038/nature10361; 10.1038/nature10361

Rasmussen, S. G., DeVree, B. T., Zou, Y., Kruse, A. C., Chung, K. Y., Kobilka, T. S., . . . Kobilka, B. K. (2011). Crystal structure of the beta2 adrenergic receptor-gs protein complex. Nature, 477(7366), 549-555. doi:10.1038/nature10361; 10.1038/nature10361

Rivero-Muller, A., Chou, Y. Y., Ji, I., Lajic, S., Hanyaloglu, A. C., Jonas, K., . . . Huhtaniemi, I. (2010). Rescue of defective G protein-coupled receptor function in vivo by intermolecular cooperation. Proceedings of the National Academy of Sciences of the United States of America, 107(5), 2319-2324. doi:10.1073/pnas.0906695106 [doi]

Smith, N. J., & Milligan, G. (2010). Allostery at G protein-coupled receptor homo- and heteromers: Uncharted pharmacological landscapes. Pharmacological Reviews, 62(4), 701-725. doi:10.1124/pr.110.002667; 10.1124/pr.110.002667

Stevens, R. C., Cherezov, V., Katritch, V., Abagyan, R., Kuhn, P., Rosen, H., & Wuthrich, K. (2013). The GPCR network: A large-scale collaboration to determine human GPCR structure and function. Nature Reviews.Drug Discovery, 12(1), 25-34. doi:10.1038/nrd3859 [doi]

UniProt Consortium. (2014). Activities at the universal protein resource (UniProt). Nucleic Acids Research, 42(Database issue), D191-8. doi:10.1093/nar/gkt1140; 10.1093/nar/gkt1140

van der Kant, R., & Vriend, G. (2014). Alpha-bulges in G protein-coupled receptors. International Journal of Molecular Sciences, 15(5), 7841-7864. doi:10.3390/ijms15057841 [doi]

Venkatakrishnan, A. J., Deupi, X., Lebon, G., Tate, C. G., Schertler, G. F., & Babu, M. M. (2013). Molecular signatures of G-protein-coupled receptors. Nature, 494(7436), 185-194. doi:10.1038/nature11896; 10.1038/nature11896

Violin, J. D., & Lefkowitz, R. J. (2007). Beta-arrestin-biased ligands at seven-transmembrane receptors. Trends in Pharmacological Sciences, 28(8), 416-422. doi:10.1016/j.tips.2007.06.006

G protein-coupled receptors: Walking hand-in-hand, talking hand-in-hand? British Journal of Pharmacology, 163(2), 246-260. doi:10.1111/j.1476-\ 5381.2011.01229.x; 10.1111/j.1476-5381.2011.01229.x

Vogler, O., Barcelo, J. M., Ribas, C., & Escriba, P. V. (2008). Membrane interactions of G proteins and other related proteins. Biochimica Et Biophysica Acta, 1778(7-8), 1640-1652. doi:10.1016/j.bbamem.2008.03.008 [doi]

Wettschureck, N., & Offermanns, S. (2005). Mammalian G proteins and their cell type specific functions. Physiological Reviews, 85(4), 1159-1204. doi:85/4/1159

Whalen, E. J., Rajagopal, S., & Lefkowitz, R. J. (2011). Therapeutic potential of beta-arrestin- and G protein-biased agonists. Trends in Molecular Medicine, 17(3), 126-139. doi:10.1016/j.molmed.2010.11.004; 10.1016/j.molmed.2010.11.004

Wootten, D., Christopoulos, A., & Sexton, P. M. (2013). Emerging paradigms in GPCR allostery: Implications for drug discovery. Nature Reviews.Drug Discovery, 12(8), 630-644. doi:10.1038/nrd4052; 10.1038/nrd4052

APPENDIX A. GLOSSARY OF TERMS RELEVANT TO G PROTEIN-COUPLED RECEPTORS

ESSENTIAL GPCR TERMS:

Ballesteros and Weinstein Nomenclature – The accepted nomenclature for referring to amino acids in the transmembrane domain of a GPCR that are assigned two numbers N1, N2. N1 being the transmembrane helix number (one through seven) and N2 being the number relative to the most conserved residue in the transmembrane helix is assigned 50 which then decreases toward the N-terminus and increases towards the C-terminus. (Moreira, 2014)

Domains – Intracellular which refers to anything inside the cell, extracellular which refers to anything outside the cell, and transmembrane which refers to the domain spanning the membrane.

GPCR Classes – GPCRs are categorized into four major classes and several subfamilies. These classes are determined based on the primary structure of the GPCR (Katritch, 2013).

Ionic Lock – A conserved motif among GPCRs. It is a bridge that keeps the interaction between R3.50 of the consensus D[E]RY in Transmembrane Helix Three with D[E]3.49 and D[E]6.30. (See Ballesteros and Weinstein Nomenclature). (Moreira, 2014)

Rotamer Toggle – An interaction among juxtaposed aromatic residues in Transmembrane Helix Six that senses the binding of the ligand and through a coordinated change in rotameric angles triggers the regulation of the ionic lock through a series of specific rearrangements in the intracellular part of the GPCR. (Moreira, 2014)

Signal Transduction – A reaction that occurs when an extracellular signaling molecule activates a cell surface receptor. This receptor then alters intracellular molecules creating a response.

Water Network – Water molecules that serve in structural and functional plasticity. These water molecules are each fixed in location buried within the interior of the transmembrane bundle, so are not part of the aqueous bulk solvent.

BASIC BIOCHEMISTRY TERMS:

Protein Structure levels of organization – There are typically four levels of organization. (Goodman & Gilman, 2010)

- **The Primary Sequence** – Refers to the amino sequence. In terms of GPCRs that would be the amino acid sequence of a polypeptide chain, presented in a linear manner starting with the N-terminus and finishing with C-terminus. Linear motifs may be presented as features of the primary sequence - for example the D[E]RY motif.
- **The Secondary Structure** – Localized short-range structural conformations created by propensity of the peptide backbone to be stabilized by hydrogen bonding. Common secondary structure motifs include: one: alpha helices, which are stabilized by backbone hydrogen bonds aligned parallel to the axis of the local structure, and two: beta strands/beta sheets which are stabilized by backbone hydrogen bonds aligned perpendicular to the local structure. Additional common secondary structure motifs include three: turns and four: coils. The prominent secondary structure motif for GPCRs is the alpha helix.
- **The Tertiary Structure** – The overall structure of a single polypeptide chain formed by folding shorter range structural domains into a unique three dimensional shape. In terms of GPCRs this refers to the "bundle of rods" that the helices form in 3D space.
- **The Quaternary Structure** – Describes the three dimensional arrangement of a protein complex comprised of multiple subunits (ie multiple polypeptide chains) defining the unique spatial relationship of the subunits to each other. In GPCRs this would refer to the formation of oligomers and receptor complex.

Amines – Organic compounds and functional groups that contain a basic nitrogen atom with a lone pair of electrons that is considered to be a derivative of ammonia. Many ligands for GPCRs are monoamines.

Effector – An effector is a target that carries out a specific event.

Cis – General term used to refer to something acting on itself. Opposite of Trans.

Trans – General term used to refer to something acing on another partner entity rather acting on itself. Opposite of Cis.

Aqueous – Solution that is contained or dissolved in water.

BASIC BIOCHEMISTRY TERMS:

Hydrophilic – Refers to a molecular entity that is attracted to and tends to be dissolved in water and other polar substances.

Hydrophobic – Refers to a molecular entity that is repelled from water. These tend to be fats or oils.

Affinity – The strength of noncovalent chemical binding between two substances as measured by the dissociation constant of the complex.

Efficacy – The measurement of the maximal response of a ligand. One hundred percent efficacy is defined by the endogenous agonist and represents the upper asymptote of a sigmoidal curve (Andresen, 2011).

Potency – The measurement of the concentration of a ligand in respect to a biological response. It is a value that indicates the ligand concentration required to reach half maximal effect. (Andresen, 2011)

Internalization – When endocytosis envelopes the GPCR and brings it into the cell. The GPCR may continue to work on the inside of the cell or it may be shuttled to lysosomes for removal by proteolytic digestion. (Luttrell, 2014).

Desensitization – Loss of responsiveness of a system despite continue presence of a stimulus. (Luttrell, 2014).

Phosphorylation – The addition of a phosphate group to a protein or other organic molecule. Phosphorylation turns many protein enzymes on and off, thereby altering their function and activity. In the context of GPCRs, GPCRs are subject to phosphorylation by G protein coupled receptor kinases (GRKS) and other kinases, and the phosphorylation of GPCRs allows them to bind to arrestins and become desensitized.

Phosphatase – An enzyme that catalyzes the hydrolysis of organic phosphates in a specified (acid or alkaline) environment. A phosphatase removes covalently bound phosphate groups from proteins, in a reaction known as dephosphorylation.

Kinase – An enzyme that catalyzes the transfer of a phosphate group from a phosphate donor to a specified molecule.

In silico – Means performed on a computer or a via computer simulation.

In vitro – Means taking place in a test tube, culture dish, or elsewhere outside a living organism.

In vivo – Means taking place in a living organism.

Steric Hindrance – Each atom within a molecule occupies a certain amount of space, and if atoms are too close together there is an associated cost in energy due.

ALLOSTERIC AND ORTHOSTERIC BINDING SITES:

Orthosteric – Ligand binding sites that are recognized by the endogenous ligand. Their locations vary based on which GPCR ligand pair is being described. If there is more than one endogenous ligand for a particular receptor site the orthosteric site is the one that binds to the major physiological ligand. Multiple ligands may bind to the site including classical agonists and classical competitive antagonists and inverse agonists. i.e. In the case of beta-adrenergic the endogenous ligand would be epinephrine so the orthosteric site would be where epinephrine binds to the GPCR. The region that is orthosteric for one GPCR may be allosteric for another GPCR (Katritch et al., 2013).

Allosteric – Interaction between two or more topographically distinct binding sites. Every binding site that is not orthosteric is allosteric (Katritch et al., 2013).

Allosteric Ligand – Ligands that bind to GPCRs at sites that are separate from the sites to which endogenous ligands binds (orthosteric). They increase the functionality of GPCRS to be manipulated for potential therapeutic benefits (Katritch et al., 2013).

Allosteric Network – Relationship between ligand binding site and effector sites is often described as allosteric because it often describes an interaction between two or more topographically distinct binding sites. (Stevens et al., 2013; Wootten et al., 2013).

TERMS RELATED TO GPCR AND LIGAND BINDING:

Ligand – A molecule that binds to a receptor or enzyme in order to elicit a chemical response. (Andresen, 2011).

TERMS RELATED TO GPCR AND LIGAND BINDING:

Ligand Binding Site – In terms of GPCRs orthosteric ligand binding for small ligands occurs within ligand binding pockets which are cavities within the transmembrane domain that generally communicate with the extracellular space. Larger ligands are accommodated within ligand binding sites formed by extracellular loops and the N-terminal extension sequences. (See Orthosteric and Allosteric). (Katritch et al., 2013).

Endogenous (natural, physiological) – Substances that originate within a system. Opposite of exogenous.

Exogenous (synthetic, surrogate, pharmacological) – An action or object that comes from outside of a system. Opposite of endogenous.

Agonist – Chemical substance that binds to a receptor of a cell and triggers a response by that cell. They often mimic the action of a naturally occurring substance. (Andresen, 2011).

Antagonist – Chemical substance that interferes with the physiological action of another, especially by combining with and blocking its receptor. (Andresen, 2011).

Functional Selectivity – Selective activation of a subset of the signaling pathways available to a receptor by a ligand.

Biased Agonist – The ability of a ligand to preferentially stabilize specific G protein-coupled receptor conformations at the exclusion of others, with each conformational state associated with its own repertoire of signaling behaviors.

Inverse Agonist – Ligands that bind to a receptor and decrease its basal signaling activity.

Partial Agonist – Agonists that produce a signaling response that is lower than the maximum response achievable for the given signaling system. Can antagonize the effects of full agonists.

Arrestins – A small family of proteins important for regulating signal transduction by activating or redirecting pathways. They play roles in receptor desensitization and internalization. Phosphorylation of the GPCR by a serine or threonine kinase allows the arrestin to bind to the GPCR and become desensitized. They were more recently recognized to elicit downstream signaling pathways that differ from those engaged by G proteins.

G Protein – Guanine nucleotide-binding proteins are a family of proteins involved in signal transduction. Their activity is regulated by factors that control their ability to bind to and hydrolyze GTP to GDP. G proteins belong to the larger group of enzymes called GTPases. There are two major subdivisions of G proteins the heterotrimeric G proteins which are the direct effectors of GPCRs, and the small G proteins activated downstream of diverse signal pathways. Heterotrimeric G proteins are comprised of three subunits a Gα, a Gβ, and a Gγ. Some important Gα's include: Gαi, Gαq/11, Gαs, Gα12/13.

Heterotrimeric G Protein – The biologically active heterotrimers G proteins that are made up alpha, beta, and gamma subunits. These are the forms of G proteins that the GPCRs typically interact with.

Orphan Receptors – An apparent receptor that has a similar structure to other identified receptors but whose endogenous ligand has not yet been identified.

THE "-MER'S"

Protomer – A structural unit of an oligomeric protein. These can be a protein subunit or several different subunits, that assemble in a defined stoichiometry to form an oligomer.

Oligomer – A macromolecular complex that consists of a few protomer subunits, formed by non-covalent bonds. (Ferre et al., 2014).

Monomer – Acting as a single unit. Monomer refers to a single chain without subunits, rather than as a part of a complex. It is often used to denote a single subunit within a complex. The term protomer may also be used

Homomer – A complex formed from a multiple identical subunits.

Heteromer – A complex formed by subunits of non-identical subunits.

Dimer – A macromolecular complex formed by two, usually non-covalently bound, macromolecules like proteins or nucleic acids. GPCR may signal in complexes of two or more.

THE "-MER'S"	**Homodimer** – A macromolecular complex formed by two identical subunits.
	Heterodimer – A macromolecular complex formed by two non-identical subunits in terms of GPCR's a heterodimer consists of a complex of two non-identical GPCRs such as the in the case of dopamine.
	Trimer – A polymer comprising three subunits.
	Heterotrimer – A macromolecular complex of three non-identical subunits.
LIPID OR LIPID-LIKE MODIFICATIONS:	**Farnesyl Group** – The parent compound Farnesol a natural 15-carbon organic compound which is an acyclic sesquiterpene alcohol with the molecular formula $C_{15}H_{26}O$. It is produced from isoprene compounds. It is added to proteins bearing a four amino acid sequence motif at the carboxyl terminus of a protein. It is a lipid-like hydrocarbon modificiation that can anchor a protein to a leaflet of a lipid bilayer. In terms of GPCRs it occurs in the G gamma protein. (Vogler, Barcelo, Ribas, & Escriba, 2008)
	Myristoyl Group – The parent compound Myristic acid is a fatty acid with the molecular formula $CH_3(CH_2)_{12}COOH$. It is highly hydrophobic and can become incorporated into the core of the phospholipid bilayer of the plasma membrane of the eukaryotic cell. It acts as an anchor in biomembranes. In terms of GPCRs it occurs in the G alpha i protein class of G proteins. (Vogler et al., 2008).
	Palmitoyl Group – The parent compound Palmitic acid is a fatty acid with the molecular formula $CH_3(CH_2)_{14}CO_2H$. It is highly hydrophobic and can become incorporated into the core of the phospholipid bilayer of the plasma membrane of the eukaryotic cell. It acts as an anchor in biomembranes. In terms of GPCRs it occurs in helix 8 of the receptor, and also in the N-terminal region of most G alpha proteins. (Chini & Parenti, 2009).
	Prenylated – The addition of hydrophobic molecules to a protein or chemical compound. It involves the transfer of either a farnesyl or a geranyl-geranyl moiety to C-terminal cysteine of the target protein. It acts as an anchor in biomembranes (Chini & Parenti, 2009).
EXPERIMENTAL AND COMPUTATIONAL METHODOLOGIES TO DETERMINE STRUCTURE AND FUNCTION	**X-Ray Crystallography** – A tool used to determine the crystal structure of an atom or molecule using x-ray that diffract in many specific directions.
	Homology Modeling – A method of generating an image of a protein without an experimentally determined structure. It can be compared to a protein that has a solved crystal structure that is similar to the protein.
	Chimera – Chimera proteins can be naturally occurring fusion proteins, or can be engineered by recombinant molecular biological techniques for purposes of generating a protein with certain desirable characteristics. In GPCR biology, engineered chimeric proteins contain segments of accessory protein sequence of the GPCR of interest, created in order to facilitate crystal formation for X-ray crystallographic structure determination.
	Nuclear Magnetic Resonance – Is an advanced medical imaging technique. It represents a physical phenomenon where the nucleus has an odd number of protons or neutrons and thus have intrinsic movement which is measured by the frequency of the resonance of the magnetic field. It is an alternate method to x-ray crystallography for determining macromolecular structure.
	FRET-Fluorescence Resonance Energy Transfer – A useful tool to quantify molecular dynamics in biophysics and biochemistry, including protein-protein interactions, protein-DNA interaction, and protein conformational changes. It uses external illumination to initiate the fluorescence It measures the nearness of two probes. If the two probes come close enough they transfer energy. The signal from the probes changes based on the nearness.
	BRET-Bioluminescence Resonance Energy Transfer – A technique (related to FRET) that uses bioluminescence luciferase to measure the nearness of two probes. If the two probes come close enough they transfer energy. This method can be preferable to –FRET which is fluorescence –based, as it avoids the need to use external illumination to initiate the fluorescence transfer which can lead to background noise in the results.

G PROTEIN SIGNALING:

GPCRs, activated by ligands, interact with a variety of heterotrimeric G Proteins. This causes the exchange of GDP for GTP bound to G protein alpha subunits followed by dissociation of the beta/gamma heterodimers. Alpha and beta/gamma subunits are active and transmit signals separately into the cells. This is shown in figure 1.

Gα_s class – The Gα_s activate all isoforms of adenylyl cyclase. Adenylyl cyclase increases levels of cyclic adenosine monophosphate (cAMP) in the cell. Elevated cAMP causes

ubiquitously or near ubiquitously. Other members of the class include Gα_{14} and G$\alpha_{15/16}$, which have limited expression patterns. (Wettschureck & Offermanns, 2005)

G$\alpha_{12/13}$ class – G$\alpha_{12/13}$ subunits interact with guanine nucleotide exchange factors to activate Rho GTPases in the cells. Rho GTPases regulate the actin cytoskeleton. They are expressed ubiquitously. (Oldham & Hamm, 2008)

The G$\beta\gamma$ subunits stay together, and tend to signal to ion channels and particular isoforms of adenylyl cyclase and phospholipase, as well as phophoinositide-3-kinase isoforms.

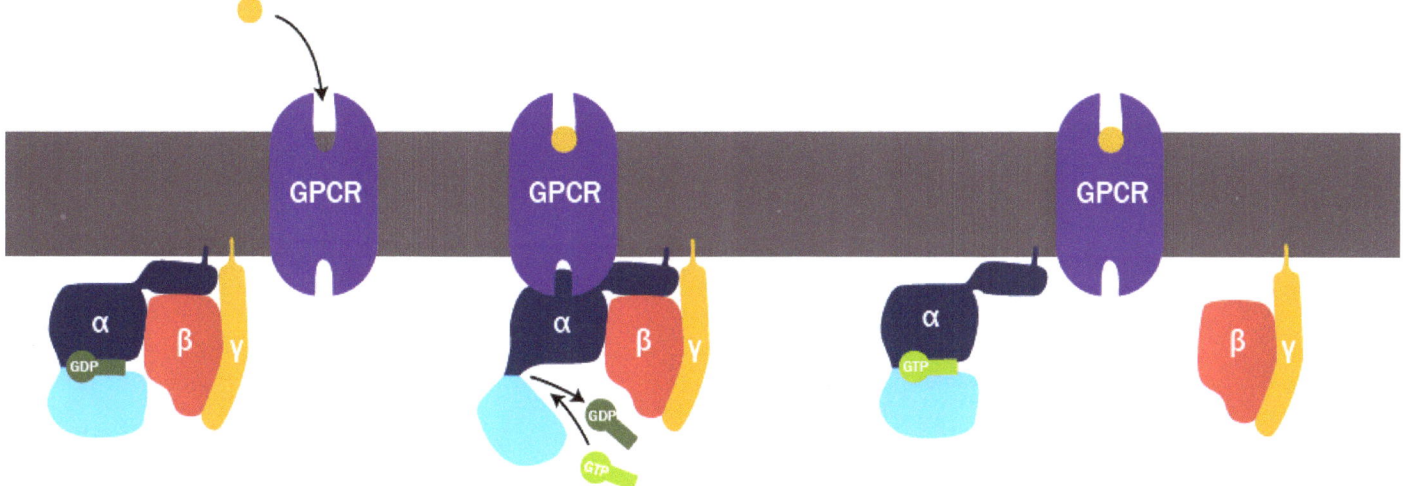

Figure 1: G Alpha protein GDP to GTP exchange bound to a GPCR (Rasmussen et al., 2011).

dissociation of the cAMP-dependent protein kinase (PKA) holenzyme complex, which results in PKA activation. PKA regulates activity of different proteins participating in various cell processes. cAMP also target small GTPases. The Gαs subunit is expressed ubiquitously. Gα_{olf} (olfactory) subunits closely related to Gα_s, are classified as members of the Gα_s class and are expressed in olfactory epithelium and brain.

G$\alpha_{i/o}$ class – The "I" in Gα_i stands for inhibitory. This is because they inhibit adenylyl cyclase to inhibit cyclase activity which decrease cAMP concentration in the cell. Gα_i subtypes are expressed either widely or ubiquitously. Gα_t (transducin) and Gα_{gust} (gustatory) subtypes are classified as members of the G$\alpha_{i/o}$ class, however rather than inhibiting andenylyl cyclase, these G Proteins activate cyclic guanosine monophosphate-specific phosphodiesterase. Gα_{t-r} is expressed in retinal rods, Gα_{t-c} is expressed in retinal cones and Gα_{gust} is expressed in taste cells. Other members of the class include Gα_o (o stands for other), and Gα_z. (Wettschureck & Offermanns, 2005)

G$\alpha_{q/11}$ class – The G$\alpha_{q/11}$ subtypes activate phospholipase C beta (PLC-beta). PLC-beta catalyzes the hydrolysis of phosphoinositide 4,5-bisphosphate to form inositol tris-phosphate (IP3) and diacylglycerol (DAG). IP3 in turn elevates cytosolic Ca2+ concentration, and DAG activates protein kinase C (PKC), which phosphorylates a number of key cellular targets. The G$\alpha_{q/11}$ subunits are expressed

There are 5 beta subunits and 14 gamma subunits that can form beta gamma complexes.

There are other regulators have been found to interact with G Proteins known as "regulators of G Protein signaling" (RGS). (Oldham & Hamm, 2008)

ARRESTINS

Arrestins are also targeted by GPCRs. There are four members of the arrestin family, two visual arrestins which have expression limited to the visual system where they interact with rhodopsins, and two non-visual arrestins with ubiquitous expression patterns. Non-visual arrestins were first identified as binding partners to the beta-adrenergic receptors, and the name beta-arrestin was assigned based on partnering to the beta adrenergic receptors, however non-visual arrestins were soon recognized to bind to GPCRs widely and are not restricted to partnering to the beta adrenergic receptors. The visual arrestins are known as Arrestin1 and Arrestin4. The two nonvisual arrestins are: Arrestin2, also known as Beta-arrestin1; and Arrestin3, also known as Beta-arrestin2. (DeWire, Ahn, Lefkowitz, & Shenoy, 2007)

The effects of arrestins are different than G proteins, i.e. they have a slower onset and a longer duration. In most settings arrestins need the phosphorylation of the GPCR by GRKs in order to bind to GPCRs, although this phosphorylation requirement is not universal. As a consequence of arrestin binding to GPCRs, arrestins initiate receptor desensitization, internalization, and G protein independent signaling events. The G protein independent signaling events include activation of non-receptor tyrosine kinases, which then activate members of the MAP kinase family; Akt signaling pathway regulation: and activation of E3 ubiquitin ligases to control ubiquitination and consequent protein degradation to name a few. Collectively arrestin functions are fundamental to receptor regulation. (DeWire et al., 2013).

Gα Signaling Pathway

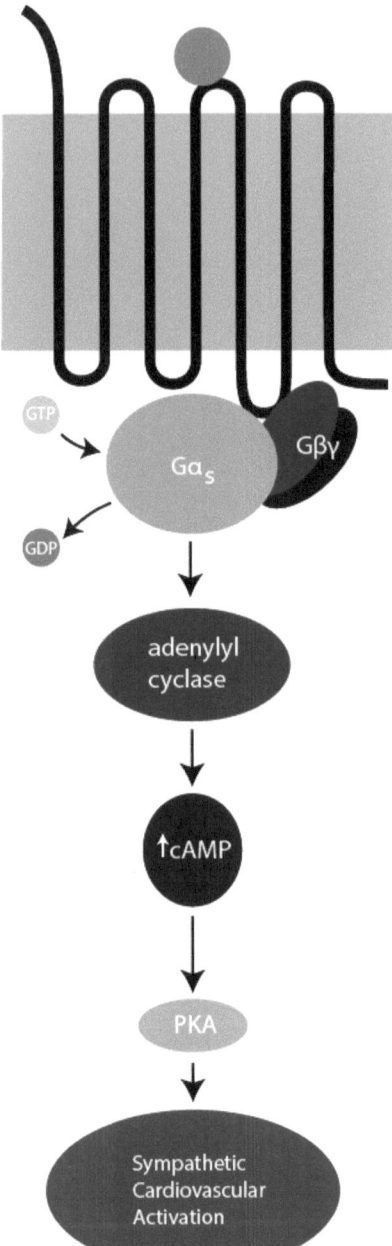

Arrestin Signaling Pathway

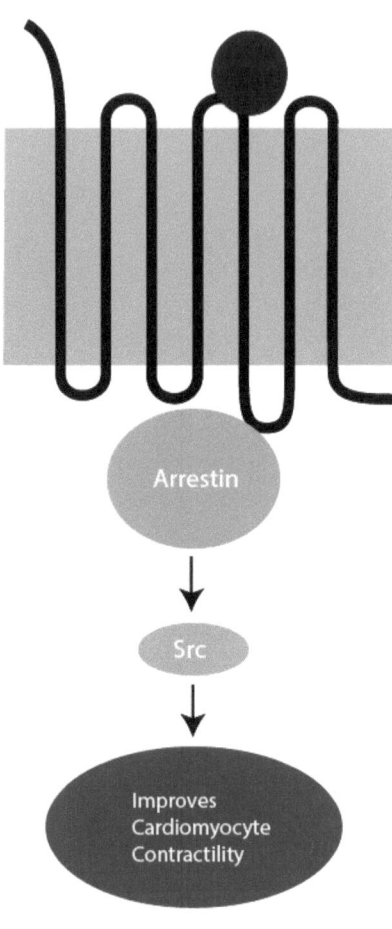

Figure 2: Typical downstream signaling cascades for a G Alpha Protein and a Beta-Arrestin

APPENDIX C. CHART OF AMINO ACIDS

ACIDIC AMINO ACIDS:

Aspartic Acid — Asp — D **Glutamic Acid — Glu — E**

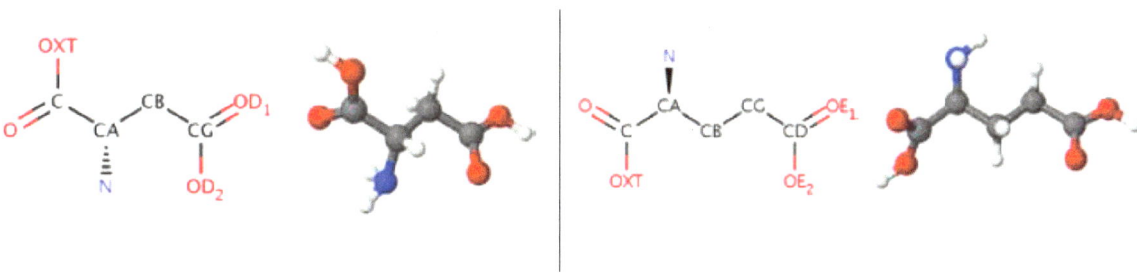

Properties: They are acidic and polar and negatively charged (anionic) at neutral pH.
D and E are acidic by virtue of the COO- carboxylic acid/carboxylate moiety that comprises part of their respective R groups.

Relevance to GPCRs: D and E are very important in GPCRs. They are the acidic component of the D[E]RY ionic lock between transmembrane helices three and six.
Highly conserved in helices two and three.

CYCLIC AMINO ACIDS:

Proline — Pro — P

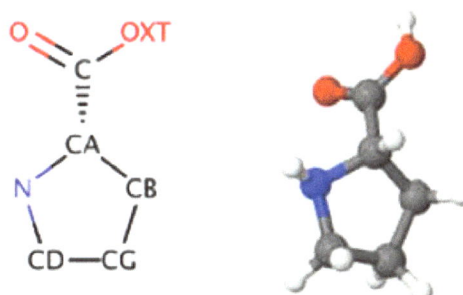

Properties: Nonpolar and uncharged at neutral pH
Usually solvent-exposed despite having an aliphatic side chain.
Contains an unusual R-group which binds back into main chain through the N-group of the main chain. This binding back into the main chain means that proline's R group is tethered, and only certain positions are possible. Proline is an "imino" acid, because the main chain N-group is branched, rather than being an "amino" acid.

Relevance to GPCRs: Helix kinks within membrane.
Part of conserved NPxxY motif in helix seven which participates in stabilizing active state and also works in the conserved motif on helix six the rotamer toggle.
Highly concerved helices two, five, six, and seven.

2D images courtesy of the Protein Data Bank retrieved Feb. 9, 2013 from: http://www.rcsb.org/pdb/home/home.do
3D images courtesy of the PubChem retrieved Feb. 9, 2013 from: https://pubchem.ncbi.nlm.nih.gov/

Vischer, H. F., Watts, A. O., Nijmeijer, S., & Leurs, R. (2011).

BASIC AMINO ACIDS:

Histidine — His — H

Properties: Basic and Polar and positively charged (cationic) at neutral pH.
Imidazole group which is a ring structure that contains basic nitrogen component.

Relevance to GPCRs: Found in Rhodopshin/opsin. Used to reveal previously inaccessible binding sites.

Arginine — Arg — R

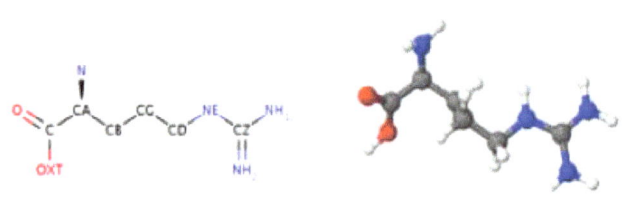

Properties: Basic and Polar and positively charged at neutral pH.
Capped by a guanidinium group which contains three nitrogen atoms.

Relevance to GPCRs: Plays a role in the ionic lock. Highly conserved in helix three and six.

Lysine — Lys — K

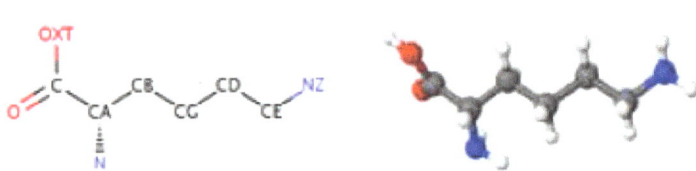

Properties: Basic and Polar and positively charged at neutral pH.
Primary amino group at epsilon position of the R group is responsible for behavior as base. Epsilon amino group can be modified by post-translational modification (acetylation, acylation, ubiquitination, sumoylation). Acetylation causes amide formation with resulting neutralization of the positive charge.

Relevance to GPCRs: Ubiquitination of lysine residues can target GCPRs. Highly conserved in helix five.

2D images courtesy of the Protein Data Bank retrieved Feb. 9, 2013 from: http://www.rcsb.org/pdb/home/home.do
3D images courtesy of the PubChem retrieved Feb. 9, 2013 from: https://pubchem.ncbi.nlm.nih.gov/

ALIPHATIC/NON-AROMATIC:

Glycine — Gly — G

Properties: Polar and uncharged at neutral pH.
Smallest and uses hydrogen as its side-chain.
Can be hydrophilic and hydrophobic.

Relevance to GPCRs: Known to functions in neurotransmitter GPCRs.

Alanine — Ala — A

Properties: Nonpolar and uncharged at neutral pH.
Small and R-group contains methyl group.

Relevance to GPCRs: Frequently used to replace other amino acids to make GPCRs more stable and easier to visualize.

Leucine — Leu — L **Isoleucine — Ile — I** **Valine — Val — V**

Properties: Nonpolar and uncharged at neutral pH.
Hydrophobic.

Relevance to GPCRs: Conserved in helix three in class A GPCR Family (I:40%, V:25%, L:11%).
L highly conserved in helix two.
Leucine-rich repeat-containing GPCR important in glycoprotein hormone receptor subclass A.

2D images courtesy of the Protein Data Bank retrieved Feb. 9, 2013 from: http://www.rcsb.org/pdb/home/home.do
3D images courtesy of the PubChem retrieved Feb. 9, 2013 from: https://pubchem.ncbi.nlm.nih.gov/

AROMATIC AMINO ACIDS:

Tyrosine — Tyr — Y

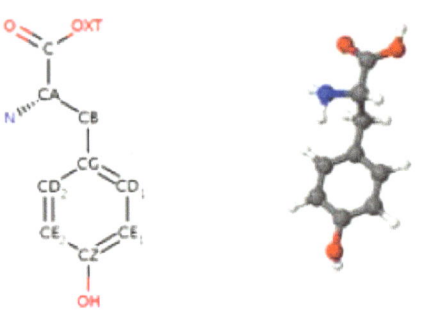

Properties: Polar and uncharged at neutral pH. Contains modified benzyl ring which is a benzene ring attached to an "OH" hydroxyl group.
The hydroxyl group is subject to post-translational modification as receiver of phosphate groups transferred by protein kinases (phosphorylation).

Relevance to GPCRs: Part of the conserved NPxxY motif in helix seven, playing a role in stabilizing helix three's DRY motif's conserved Arg3.50 through hydrogen bonding. Highly conserved motif in helices three, five, and seven.

Phenylalanine — Phe — F

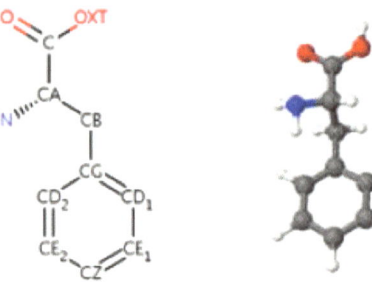

Properties: Nonpolar and uncharged at neutral pH. Benzyl side chain. Hydrophobic.

Relevance to GPCRs: Conserved in helix six contributes to outward movement and rotation of helix six.

Tryptophan — Trp — W

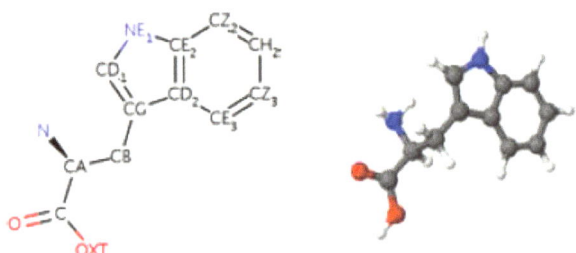

Properties: Nonpolar and uncharged at neutral pH
Contains an indole functional group which is an aromatic heterocyclic organic compound which includes a six-membered benzene ring fused to a five-membered nitrogen-containing pyrrole ring.

Relevance to GPCRs: Plays a role in the rotamer toggle switch
Highly conserved in helices three and six

2D images courtesy of the Protein Data Bank retrieved Feb. 9, 2013 from: http://www.rcsb.org/pdb/home/home.do
3D images courtesy of the PubChem retrieved Feb. 9, 2013 from: https://pubchem.ncbi.nlm.nih.gov/

ACYCLIC HYDROXYL OR SULFUR-CONTAINING AMINO ACIDS:

Serine — Ser — S

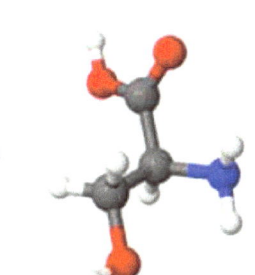

Threonine — Thr — T

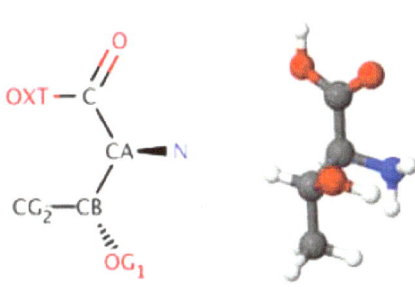

Properties: Polar and uncharged at neutral pH.
Contains "OH" hydroxyl group (alcohol).
The hydroxyl group is subject to post-translational modification, serving as the recipient of phosphate groups transferred by protein kinases (phosphorylation).

Relevance to GPCRs: Is phosphorylated by G protein-coupled receptor kinases.
The phosphorylated version serves as a binding site for arrestins, which prevent reasssociation of G proteins with their receptors.
S Highly conserved in helix five.

Asparagine — Asn — N

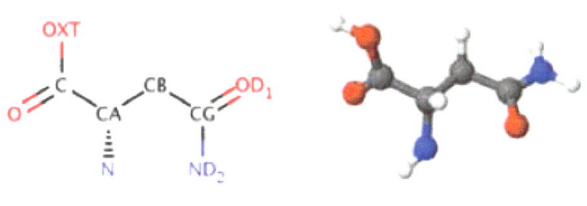

Properties: Polar and uncharged at neutral pH
Has an amide group at the terminus of the R group
Closely related structurally to aspartic acid

Relevance to GPCRs: Part of conserved motif in helix seven: NPxxY – this motif plays a role in stabilizing active state (see tyrosine entry).
Highly conserved amino acid in helix one and seven.

Glutamine — Gln — Q

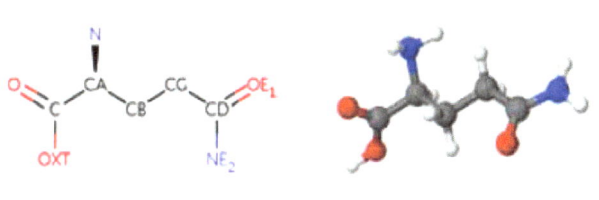

Properties: Polar and uncharged at neutral pH.
Has an amide side-chain.
Closely related structurally to glutamic acid.

Relevance to GPCRs: Conserved in helix four.

2D images courtesy of the Protein Data Bank retrieved Feb. 9, 2013 from: http://www.rcsb.org/pdb/home/home.do
3D images courtesy of the PubChem retrieved Feb. 9, 2013 from: https://pubchem.ncbi.nlm.nih.gov/

ACYCLIC HYDROXYL OR SULFUR-CONTAINING AMINO ACIDS:

Cysteine — Cys — C

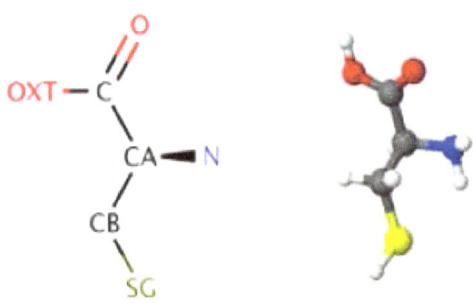

Properties: Nonpolar and uncharged at neutral pH
Forms disulfide bridge bond
Only amino acid whose side chain can form covalent bonds. The oxydized form is covalently bonded as opposed to the reduced form where each cysteine would be in its own separate form.

Relevance to GPCRs: Plays a role in the rotamer toggle switch (a conserved structural motif).
Conserved cysteine bridge between extracellular loop two and the top of transmembrane helix three helps hold helix three in it unusual diagonal angle uncharged at neutral pH.
Conserved motif in helix three and six.

Methionine — Met — M

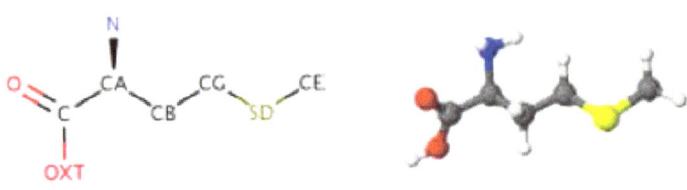

Properties: Nonpolar and uncharged at neutral pH.
Contains sulfur side – hydrophobic.
Always the first amino acid in the start codon chain in any protein including GPCRs.

Relevance to GPCRs: Conserved in helix three as a part of hydrophobic amino acids form ionic cage.

2D images courtesy of the Protein Data Bank retrieved Feb. 9, 2013 from: http://www.rcsb.org/pdb/home/home.do
3D images courtesy of the PubChem retrieved Feb. 9, 2013 from: https://pubchem.ncbi.nlm.nih.gov/

www.ingramcontent.com/pod-product-compliance
Lightning Source LLC
Chambersburg PA
CBHW051111180526
45172CB00002B/862